Mathuranathan Viswanathan

Digital Modulations using Python

December 2019

Copyright © 2019 Mathuranathan Viswanathan. All rights reserved.

Copyright © 2019 Mathuranathan Viswanathan

Cover design © 2019 Mathuranathan Viswanathan

Cover art © 2019 Mathuranathan Viswanathan

All rights reserved.

No part of this publication may be reproduced, distributed, or transmitted in any form or by any means, including photocopying, recording, or other electronic or mechanical methods, without the prior written permission of the author, except in the case of brief quotations embodied in critical reviews and certain other noncommercial uses permitted by copyright law. For permission requests, write to

mathuranathan@gmail.com

https://www.gaussianwaves.com

ISBN: 9781712342749

Black & white edition

First published 2019

The author has used his best endeavours to ensure that the URLs for external websites referred in this book are correct and active at the time of publishing. However, the author bears no responsibility for the referred websites, and can make no guarantee that a site will remain live or that the content is or will remain appropriate.

Python logo featured on the front cover is a trademark of the Python Software Foundation.

Dedicated to *Advaith*

Preface

Introduction

There exist many textbooks that provide an in-depth treatment of various topics in digital modulation techniques. Most of them underscore different theoretical aspects of design and performance analysis of digital modulation techniques. Only a handful of books provide insight on how these techniques can be modeled and simulated. Predominantly, such books utilize the sophisticated built-in functions or toolboxes that are already available in software like Matlab. These built-in functions or toolboxes hide a lot of background computations from the user thereby making it difficult, especially for a learner, to understand how certain techniques are actually implemented inside those functions.

In this book, I intend to show how the theoretical aspects of a digital modulation-demodulation system can be translated into simulation models, using existing packages in Python3 (Python version 3).

Why Python ?

Python is a free, open source, high level general purpose language that comes with a rich repertoire of robust standard libraries. As a free open source language, python helps us to reduce the development costs significantly. The code written in python is easily readable by humans and is compatible with major platforms and operating systems. Python distributions are also available for Arduino microcontroller and Rasberry Pi minicomputer.

Python comes with a range of packages like SciPy and NumPy, that support complex numerical and scientific computing. Above all, python is supported by a very strong world-wide developer community that are instrumental in providing thousands of packages for free, which enable us to accomplish complex development tasks.

Python has maintained the top spot, since 2017, in the IEEE Spectrum's annual interactive ranking of the top programming languages.

Intended audience

I intend the book to be used primarily by undergraduate and graduate students in electrical engineering discipline, who wish to learn the basic implementation aspects of a modulation demodulation technique. I assume that the reader has a fair understanding on the fundamentals of programming in Python. Readers may consult other online documentations that cover those topics.

Organization of topics

Theoretical aspects of digital modulation techniques will be kept brief. For each topic discussed, a short theoretical background is provided along with the implementation details in the form of Python3 scripts or functions. The codes written in Python carry inline comments intended to help the reader understand the flow of implementation.

As for the topics concerned, only the basic techniques of modulation and demodulation of various digital modulation techniques are covered. Waveform simulation technique and the complex equivalent baseband simulation model will be provided on a case-by-case basis. Performance simulations of well known digital modulation techniques are also provided. Additionally, simulation and performance of receiver impairments are also provided in a separate chapter.

Chapter 1 introduces some of the basic signal processing concepts that will be used throughout this book. Concepts covered in this chapter include- signal generation techniques for generating well known test signals, interpreting FFT results and extracting magnitude/phase information using FFT, computation of power and energy of a signal, various methods to compute convolution of two signals. Chapter 2 covers the waveform simulation technique for the following digital modulations: BPSK, differentially encoded BPSK, differential BPSK, QPSK, offset QPSK, pi/4 QPSK, CPM and MSK, GMSK, FSK. Power spectral density (PSD) and performance analysis for these techniques are also provided. Chapter 3 covers the complex baseband equivalent models for techniques such as M-ary PAM, M-ary PSK, M-ary QAM and M-ary FSK modulations. Chapter 4 covers the performance simulation using the models built in Chapter 3. Chapter 5 covers the aspects of using various linear equalizers in a simple communication link. Design and implementation of two important types of equalizers namely the zero-forcing equalizer and the MMSE equalizer are covered. Chapter 6 covers the topic of modeling receiver impairment, estimation and compensation for such impairments and a sample performance simulation for such case. Reference texts are cited in square brackets within the text and the references are provided at the end of each chapter.

Code documentation

Package download and documentation for the Python3 functions/scripts shown in this book, is available at the following URL. The scripts shown in this book are packaged and named as `DigiCommPy`. Make sure to include the package directory in the PYTHONPATH environmental variable.

https://www.gaussianwaves.com/DigiCommPy

The scripts are thoroughly checked for integrity and they will execute without any error. If you face any issues during execution, do not hesitate to provide feedback or contact me via the email provided below.

Acknowledgments

Finally, it is a pleasure to acknowledge the help I received while writing this book. I thank my wife *Varsha Mathuranathan* for getting the manuscript edited so quickly and for her support during the review process that greatly helped improve the manuscript. I also thank the numerous reviewers for their generous comments that helped improve the contents of this book.

Singapore,
December 2019

Mathuranathan Viswanathan
mathuranathan@gmail.com

Contents

1 Essentials of Signal Processing 1
  1.1 Generating standard test signals 1
      1.1.1 Sinusoidal signals 1
      1.1.2 Square wave 3
      1.1.3 Rectangular pulse 5
      1.1.4 Gaussian pulse 6
      1.1.5 Chirp signal 7
  1.2 Interpreting FFT results - complex DFT, frequency bins and FFTShift 10
      1.2.1 Real and complex DFT 10
      1.2.2 Fast Fourier Transform (FFT) 12
      1.2.3 Interpreting the FFT results 13
      1.2.4 FFTShift 15
      1.2.5 IFFTShift 17
      1.2.6 Some observations on FFTShift and IFFTShift 18
  1.3 Obtaining magnitude and phase information from FFT 19
      1.3.1 Discrete-time domain representation 19
      1.3.2 Representing the signal in frequency domain using FFT 19
      1.3.3 Reconstructing the time domain signal from the frequency domain samples 22
  1.4 Power spectral density 23
  1.5 Power and energy of a signal 25
      1.5.1 Energy of a signal 25
      1.5.2 Power of a signal 26
      1.5.3 Classification of signals 27
      1.5.4 Computation of power of a signal - simulation and verification 27
  1.6 Polynomials, convolution and Toeplitz matrices 30
      1.6.1 Polynomial functions 30
      1.6.2 Representing single variable polynomial functions 31
      1.6.3 Multiplication of polynomials and linear convolution 31
      1.6.4 Toeplitz matrix and convolution 32
  1.7 Methods to compute convolution 33
      1.7.1 Method 1: Brute-force method 34
      1.7.2 Method 2: Using Toeplitz matrix 34
      1.7.3 Method 3: Using FFT to compute convolution 35
      1.7.4 Miscellaneous methods 37
  1.8 Analytic signal and its applications 37
      1.8.1 Analytic signal and Fourier transform 37

		1.8.2 Applications of analytic signal 42
	1.9	Choosing a filter : FIR or IIR : understanding the design perspective 47
		1.9.1 Design specification 48
		1.9.2 General considerations in design 48
	References 54

2 Digital Modulators and Demodulators - Passband Simulation Models 55
2.1 Introduction 55
2.2 Binary Phase Shift Keying (BPSK) 56
    2.2.1 BPSK transmitter 56
    2.2.2 BPSK receiver 57
    2.2.3 End-to-end simulation 58
2.3 Coherent detection of Differentially Encoded BPSK (DEBPSK) 59
2.4 Differential BPSK (D-BPSK) 62
    2.4.1 Sub-optimum receiver for DBPSK 63
    2.4.2 Optimum non-coherent receiver for DBPSK 64
2.5 Quadrature Phase Shift Keying (QPSK) 67
    2.5.1 QPSK transmitter 68
    2.5.2 QPSK receiver 70
    2.5.3 Performance simulation over AWGN 72
2.6 Offset QPSK (O-QPSK) 73
2.7 $\pi/4$-DQPSK 79
2.8 Continuous Phase Modulation (CPM) 85
    2.8.1 Motivation behind CPM 85
    2.8.2 Continuous Phase Frequency Shift Keying (CPFSK) modulation 85
    2.8.3 Minimum Shift Keying (MSK) 87
2.9 Investigating phase transition properties 94
2.10 Power spectral density (PSD) plots 98
2.11 Gaussian Minimum Shift Keying (GMSK) 100
    2.11.1 Pre-modulation Gaussian low pass filter 100
    2.11.2 Quadrature implementation of GMSK modulator 102
    2.11.3 GMSK spectra 105
    2.11.4 GMSK demodulator 105
    2.11.5 Performance 107
2.12 Frequency Shift Keying (FSK) 109
    2.12.1 Binary-FSK (BFSK) 109
    2.12.2 Orthogonality condition for non-coherent BFSK detection 109
    2.12.3 Orthogonality condition for coherent BFSK 111
    2.12.4 Modulator 111
    2.12.5 Coherent demodulator 113
    2.12.6 Non-coherent demodulator 114
    2.12.7 Performance simulation 116
    2.12.8 Power spectral density 117
References 118

3 Digital Modulators and Demodulators - Complex Baseband Equivalent Models 121
3.1 Introduction 121
3.2 Complex baseband representation of modulated signal 121
3.3 Complex baseband representation of channel response 122
3.4 Implementing complex baseband modems using object oriented programming 123
    3.4.1 Pulse Amplitude Modulation (M-PAM) modem 125

Contents

      3.4.2   Phase Shift Keying Modulation (M-PSK) modem 126
      3.4.3   Quadrature Amplitude Modulation (M-QAM) modem 127
      3.4.4   Optimum detector on IQ plane using minimum Euclidean distance 130
      3.4.5   M-ary Frequency Shift Keying modem 131
  3.5   Instantiation of modems 135
  References 138

4 Performance of Digital Modulations over Wireless Channels 139
  4.1   AWGN channel 139
      4.1.1   Signal to noise ratio (SNR) definitions 139
      4.1.2   AWGN channel model 140
      4.1.3   Theoretical symbol error rates 142
      4.1.4   Unified simulation model for performance simulation 144
  4.2   Fading channels 146
      4.2.1   Linear time invariant channel model and FIR filters 147
      4.2.2   Simulation model for detection in flat fading channel 148
      4.2.3   Rayleigh flat-fading channel 149
      4.2.4   Rician flat-fading channel 154
  References 158

5 Linear Equalizers 159
  5.1   Introduction 159
  5.2   Linear equalizers 160
  5.3   Symbol-spaced linear equalizer channel model 162
  5.4   Implementing equalizers using object oriented programming 163
  5.5   Zero-forcing equalizer 165
      5.5.1   Least squares solution 166
      5.5.2   Noise enhancement 167
      5.5.3   Design and simulation of zero-forcing equalizer 168
      5.5.4   Drawbacks of zero-forcing equalizer 173
  5.6   Minimum mean square error (MMSE) equalizer 174
      5.6.1   Alternate solution 176
      5.6.2   Design and simulation of MMSE equalizer 177
  5.7   Equalizer delay optimization 179
  5.8   BPSK modulation with zero-forcing and MMSE equalizers 181
  5.9   Adaptive equalizer: Least mean square (LMS) algorithm 185
  References 187

6 Receiver Impairments and Compensation 189
  6.1   Introduction 189
  6.2   DC offsets and compensation 192
  6.3   IQ imbalance model 193
  6.4   IQ imbalance estimation and compensation 194
      6.4.1   Blind estimation and compensation 194
      6.4.2   Pilot based estimation and compensation 195
  6.5   Visualizing the effect of receiver impairments 197
  6.6   Performance of M-QAM modulation with receiver impairments 198
  References 202

Index 205

Chapter 1
Essentials of Signal Processing

Abstract This chapter introduces some of the basic signal processing concepts that will be used throughout this book. The goal is to enable the reader to appreciate the concepts and apply them in building a basic communication system. Concepts covered include - signal generation techniques for generating well known test signals like rectangular pulse, sine wave, square wave, chirp signal and gaussian pulse, interpreting FFT results and extracting magnitude/phase information using FFT, computation of power and energy of a signal, various methods to compute convolution of two signals, analytic signal and applications, FIR/IIR filters.

1.1 Generating standard test signals

In experimental modeling and simulation, simple test inputs such as sinusoidal, rectangular pulse, gaussian pulse, and chirp signals are widely used. These test signals act as stimuli for the simulation model and the response of the model to the stimuli is of great interest in design verification. These signals can be generated by writing our own functions or by utilizing SciPy Python library.

1.1.1 Sinusoidal signals

In order to generate a sine wave, the first step is to fix the frequency f of the sine wave. For example, we wish to generate a $f = 10 Hz$ sine wave whose minimum and maximum amplitudes are $-1V$ and $+1V$ respectively. Given the frequency of the sinewave, the next step is to determine the sampling rate.

For baseband signals, the sampling is straight forward. By *Nyquist Shannon sampling theorem*, for faithful reproduction of a continuous signal in discrete domain, one has to sample the signal at a rate f_s higher than at-least twice the maximum frequency f_m contained in the signal (actually, it is twice the one-sided bandwidth occupied by a real signal. For a baseband signal bandwidth (0 to f_m) and maximum frequency f_m in a given band are equivalent).

Let us write a function to generate a sinusoidal signal using the Python's Numpy library. Numpy is a fundamental library for scientific computations in Python. In order to use the numpy package, it needs to be imported. Here, we are importing the numpy package and renaming it as a shorter alias np.

```
import numpy as np
```

Next, we define a function for generating a sine wave signal with the required parameters.

Program 1: DigiCommPy\signalgen.py: Function to generate sine wave

```python
def sine_wave(f,overSampRate,phase,nCyl):
    """
    Generate sine wave signal with the following parameters
    Parameters:
        f : frequency of sine wave in Hertz
        overSampRate : oversampling rate (integer)
        phase : desired phase shift in radians
        nCyl : number of cycles of sine wave to generate
    Returns:
        (t,g) : time base (t) and the signal g(t) as tuple
    Example:
        f=10; overSampRate=30;
        phase = 1/3*np.pi;nCyl = 5;
        (t,g) = sine_wave(f,overSampRate,phase,nCyl)
    """
    fs = overSampRate*f # sampling frequency
    t = np.arange(0,nCyl*1/f-1/fs,1/fs) # time base
    g = np.sin(2*np.pi*f*t+phase) # replace with cos if a cosine wave is desired
    return (t,g) # return time base and signal g(t) as tuple
```

We note that the function *sine_wave* is defined inside a file named *signalgen.py*. We will add more such similar functions in the same file. The intent is to hold all the related signal generation functions, in a single file. This approach can be extended to object oriented programming. Now that we have defined the *sine_wave* function in *signalgen.py*, all we need to do is call it with required parameters and plot the output.

Program 2: DigiCommPy\chapter_1\demo_scripts.py: Call the function and plot output

```python
def sine_wave_demo():
    """
    Simulate a sinusoidal signal with given sampling rate
    """
    import numpy as np
    import matplotlib.pyplot as plt # library for plotting
    from DigiCommPy.signalgen import sine_wave # import the function
    
    f = 10 #frequency = 10 Hz
    overSampRate = 30 #oversammpling rate
    phase = 1/3*np.pi #phase shift in radians
    nCyl = 5 # desired number of cycles of the sine wave
    (t,g) = sine_wave(f,overSampRate,phase,nCyl) #function call
    
    plt.plot(t,g) # plot using pyplot library from matplotlib package
    plt.title('Sine wave f='+str(f)+' Hz') # plot title
    plt.xlabel('Time (s)') # x-axis label
    plt.ylabel('Amplitude') # y-axis label
    plt.show() # display the figure
```

Python is an interpreter based software language that processes everything in digital. In order to obtain a smooth sine wave, the sampling rate must be far higher than the prescribed minimum required sampling rate, that is at least twice the frequency f - as per *Nyquist-Shannon theorem*. Hence, we need to sample the input

1.1 Generating standard test signals

signal at a rate significantly higher than what the Nyquist criterion dictates. Higher oversampling rate requires more memory for signal storage. It is advisable to keep the oversampling factor to an acceptable value.

An oversampling factor of 30 is chosen in the previous function. This is to plot a smooth continuous-like sine wave. Thus, the sampling rate becomes $f_s = 30 \times f = 30 \times 10 = 300 Hz$. If a phase shift is desired for the sine wave, specify it too.

In Python, the function can be quickly loaded and executed with the following expressions in the command line. The resulting plot is given in Figure 1.1.

```
>> import sys
>> sys.path.append('path_where_digicommpy_resides') #search path
>> from DigiCommPy.chapter_1.demo_scripts import sine_wave_demo #import function
>> sine_wave_demo() #call the function to execute it
```

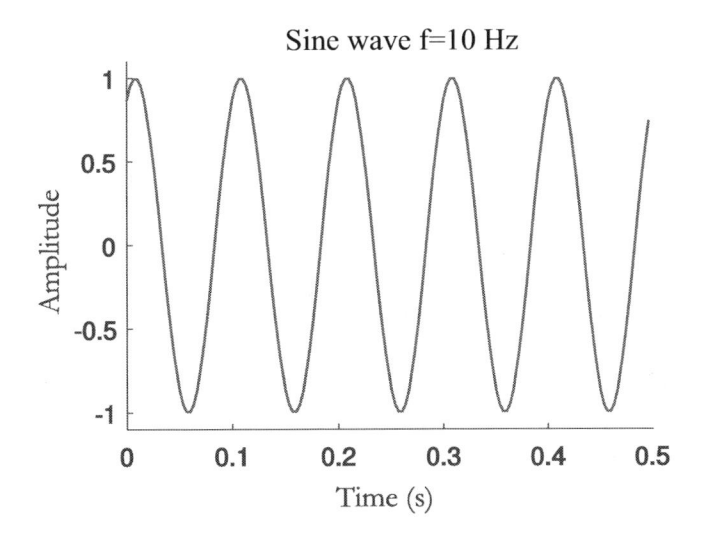

Fig. 1.1: A $10Hz$ sinusoidal wave with 5 cycles and phase shift $1/3\pi$ radians

1.1.2 Square wave

The most logical way of transmitting information across a communication channel is through a stream of square pulse – a distinct pulse for '0' and another for '1'. Digital signals are graphically represented as square waves with certain symbol/bit period. Square waves are also used universally in switching circuits, as clock signals synchronizing various blocks of digital circuits, as reference clock for a given system domain and so on.

Square wave manifests itself as a wide range of harmonics in frequency domain and therefore can cause electromagnetic interference. Square waves are periodic and contain odd harmonics when expanded as Fourier Series (where as signals like saw-tooth and other real word signals contain harmonics at all integer frequencies). Since a square wave literally expands to infinite number of odd harmonic terms in frequency domain, approximation of square wave is another area of interest. The number of terms of its Fourier Series expansion, taken for approximating the square wave is often seen as *Gibbs phenomenon*, which manifests as ringing effect at the corners of the square wave in time domain.

True square waves are a special class of rectangular waves with 50% duty cycle. Varying the duty cycle of a rectangular wave leads to pulse width modulation, where the information is conveyed by changing the duty-cycle of each transmitted rectangular wave. A true square wave can be simply generated by applying *signum function* numpy.sign over a periodic function.

$$g(t) = sgn\left[sin(2\pi ft)\right] \quad (1.1)$$

where, f is the desired frequency of the square wave and the signum function is defined as

$$sgn(x) = \begin{cases} -1 & if\ x < 0, \\ 0 & if\ x = 0, \\ 1 & if\ x > 0 \end{cases} \quad (1.2)$$

Program 3: DigiCommPy\signalgen.py: Function to generate a square wave

```
def square_wave(f,overSampRate,nCyl):
    """
    Generate square wave signal with the following parameters
    Parameters:
        f : frequency of square wave in Hertz
        overSampRate : oversampling rate (integer)
        nCyl : number of cycles of square wave to generate
    Returns:
        (t,g) : time base (t) and the signal g(t) as tuple
    Example:
        f=10; overSampRate=30;nCyl = 5;
        (t,g) = square_wave(f,overSampRate,nCyl)
    """
    fs = overSampRate*f # sampling frequency
    t = np.arange(0,nCyl*1/f-1/fs,1/fs) # time base
    g = np.sign(np.sin(2*np.pi*f*t)) # replace with cos if a cosine wave is desired
    
    return (t,g) # return time base and signal g(t) as tuple
```

We can also utilize the inbuilt *square* function from the SciPy library, for generating a periodic square waveform with a given duty cycle. The sample code, shown next, is for generating a $f = 10Hz$ square wave with 20% duty cycle.

Program 4: DigiCommPy\chapter_1\demo_scripts.py: Generate square wave using SciPy

```
def scipy_square_wave():
    """
    Generate a square wave with given sampling rate
    """
    import numpy as np
    import matplotlib.pyplot as plt
    from scipy import signal
    
    f = 10 # f = 10Hz
```

1.1 Generating standard test signals

```
overSampRate = 30 # oversampling rate
nCyl = 5 # number of cycles to generate

fs = overSampRate*f # sampling frequency
t = np.arange(start=0,stop=nCyl*1/f,step=1/fs)# time base
g = signal.square(2 * np.pi * f * t, duty = 0.2)
plt.plot(t,g); plt.show()
```

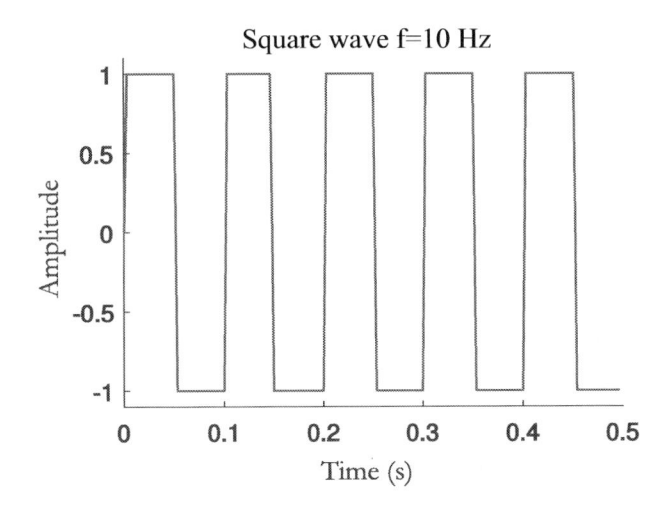

Fig. 1.2: A $10Hz$ square wave with 5 cycles and 50% duty cycle

1.1.3 Rectangular pulse

An isolated rectangular pulse of amplitude A and duration T is represented mathematically as

$$g(t) = A \cdot rect\left(\frac{t}{T}\right) \qquad (1.3)$$

where,

$$rect(t) = \begin{cases} 1 & if\ |t| < \frac{1}{2} \\ \frac{1}{2} & if\ |t| = \frac{1}{2} \\ 0 & if\ |t| > \frac{1}{2} \end{cases} \qquad (1.4)$$

The following function simulates a rectangular pulse with desired pulse width and the resulting plot is shown in Figure 1.3.

Program 5: DigiCommPy\signalgen.py: Generating rectangular pulse with desired pulse width

```
def rect_pulse(A,fs,T):
    """
    Generate isolated rectangular pulse with the following parameters
```

```
    Parameters:
        A : amplitude of the rectangular pulse
        fs : sampling frequency in Hz
        T : duration of the pulse in seconds
    Returns:
        (t,g) : time base (t) and the signal g(t) as tuple
    Example:
        A = 1; fs=500;T = 0.2;
        (t,g) = rect_pulse(A,fs,T)
    """
    t = np.arange(-0.5,0.5,1/fs) # time base
    rect = (t >-T/2) * (t<T/2) + 0.5*(t==T/2) + 0.5*(t==-T/2)
    g = A*rect
    return (t,g)
```

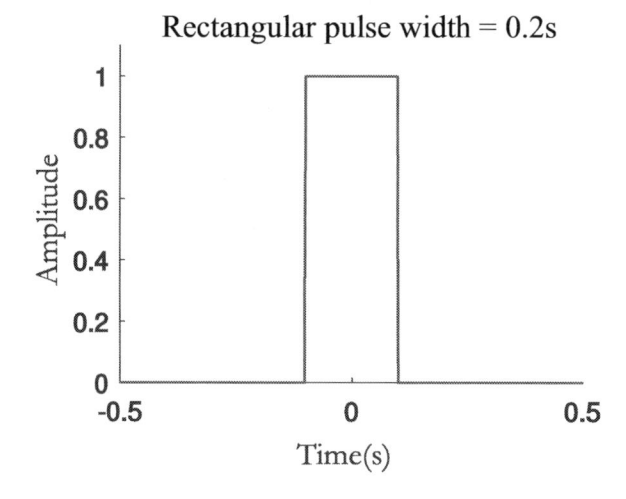

Fig. 1.3: A rectangular pulse having pulse-width 0.2*s*

1.1.4 Gaussian pulse

In digital communications, Gaussian filters are employed in *Gaussian Minimum Shift Keying - GMSK* (see section 2.11) and *Gaussian Frequency Shift Keying (GFSK)*. Two dimensional Gaussian filters are used in image processing for producing Gaussian blurs. The impulse response of a Gaussian filter is Gaussian. Gaussian filters give no overshoot with minimal rise and fall time, when excited with a step function. Gaussian filter has minimum group delay. The impulse response of a Gaussian filter is written as a Gaussian function

$$g(t) = \frac{1}{\sqrt{2\pi}\sigma}e^{-\frac{t^2}{2\sigma^2}} \tag{1.5}$$

The following function generates a Gaussian pulse with $\sigma = 0.1s$. The resulting plot is given in Figure 1.4

1.1 Generating standard test signals

Program 6: DigiCommPy\signalgen.py: Generating Gaussian pulse with desired pulse width

```python
def gaussian_pulse(fs,sigma):
    """
    Generate isolated Gaussian pulse with the following parameters
    Parameters:
        fs : sampling frequency in Hz
        sigma : pulse width in seconds
    Returns:
        (t,g) : time base (t) and the signal g(t) as tuple
    Example:
        fs = 80; sigma = 0.1;
        (t,g) = gaussian_pulse(fs,sigma)
    """
    t = np.arange(-0.5,0.5,1/fs) # time base
    g = 1/(np.sqrt(2*np.pi)*sigma)*(np.exp(-t**2/(2*sigma**2)))
    return(t,g) # return time base and signal g(t) as tuple
```

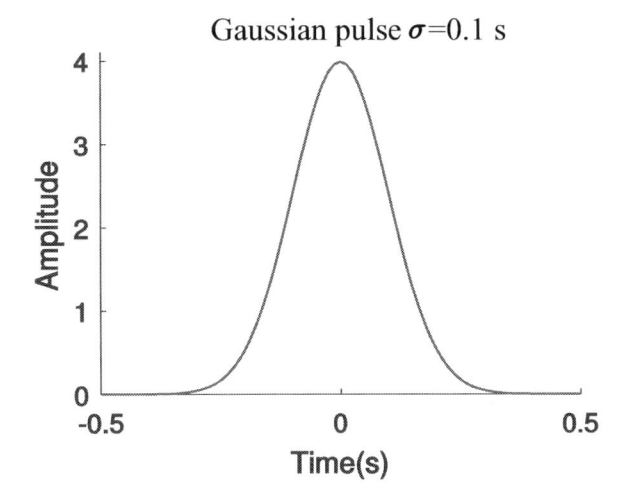

Fig. 1.4: A Gaussian pulse with $\sigma = 0.1s$

1.1.5 Chirp signal

All the signals, discussed so far, do not change in frequency over time. Obtaining a signal with time-varying frequency is of main focus here. A signal that varies in frequency over time is called *chirp*. The frequency of the chirp signal can vary from low to high frequency (up-chirp) or from high to low frequency (low-chirp).

Chirp signals are encountered in many applications ranging from radar, sonar, spread spectrum, optical communication, image processing, doppler effect, motion of a pendulum, as gravitation waves, manifestation as frequency modulation (FM), echo location etc.

A linear chirp signal sweeps the frequency from low to high frequency (or vice-versa) linearly. One approach to generate a chirp signal is to concatenate a series of segments of sine waves each with increasing (or decreasing) frequency in order. This method introduces discontinuities in the chirp signal due to the mismatch in the phases of each such segments. Modifying the equation of a sinusoid to generate a chirp signal is a better approach.

The equation for generating a cosine signal with amplitude A, angular frequency ω_0 and initial phase ϕ is

$$x(t) = A\,cos(\omega_0 t + \phi) \tag{1.6}$$

This can be written as a function of instantaneous phase

$$x(t) = A\,cos[\theta(t)] \tag{1.7}$$

where $\theta(t) = \omega_0 t + \phi$ is the instantaneous phase of the sinusoid and it is *linear in time*. The time derivative of instantaneous phase $\theta(t)$, is equal to the angular frequency ω_0 of the sinusoid.

$$\frac{d}{dt}\theta(t) = \omega_0 \tag{1.8}$$

Instead of having the phase linear in time, let's change the phase to quadratic form and thus render it non-linear. For some constant α,

$$\theta(t) = 2\pi\alpha t^2 + 2\pi f_0 t + \phi \tag{1.9}$$

Therefore, the equation for chirp signal takes the following form,

$$x(t) = A\,cos\left(\theta(t)\right) = A\,cos\left(2\pi\alpha t^2 + 2\pi f_0 t + \phi\right) \tag{1.10}$$

The first derivative of the phase is the instantaneous angular frequency as given by

$$\omega_i(t) = \frac{d}{dt}\theta(t) = 4\pi\alpha t + 2\pi f_0 \tag{1.11}$$

Hence, the time-varying frequency in Hz is given by

$$f_i(t) = 2\alpha t + f_0 \tag{1.12}$$

In the above equation, the frequency is no longer a constant, rather it is of time-varying nature with initial frequency given by f_0. Thus, from the above equation, given a time duration T, the rate of change of frequency is given by

$$k = 2\alpha = \frac{f_1 - f_0}{T} \tag{1.13}$$

where, f_0 is the starting frequency of the sweep, f_1 is the frequency at the end of the duration T. Substituting equations 1.12 and 1.13 in 1.11,

$$\omega_i(t) = \frac{d}{dt}\theta(t) = 2\pi(kt + f_0) \tag{1.14}$$

From equations 1.11 and 1.13

$$\theta(t) = \int \omega_i(t)\,dt$$

$$= 2\pi \int (kt + f_0)\,dt = 2\pi\left(k\frac{t^2}{2} + f_0 t\right) + \phi_0$$

$$= 2\pi\left(k\frac{t^2}{2} + f_0 t\right) + \phi_0 = 2\pi\left(\frac{k}{2}t + f_0\right)t + \phi_0 \tag{1.15}$$

1.1 Generating standard test signals

where, ϕ_0 is a constant which will act as the initial phase of the sweep. Thus the modified equation for generating a chirp signal (from equations 1.10 and 1.15) is given by

$$x(t) = A\,cos\,(\theta(t)) = A\,cos\,[2\pi f(t)\,t + \phi_0] \quad (1.16)$$

where, the time-varying frequency function is given by

$$f(t) = \frac{k}{2}t + f_0 \quad (1.17)$$

A chirp signal can be easily generated using the Scipy's chirp function. The following script utilizes the Scipy library, generates a chirp with starting frequency $f_0 = 1Hz$ at the start of the time base and $f_1 = 20Hz$ at $t_1 = 0.5s$. The resulting plot is shown in Figure 1.5.

Program 7: DigiCommPy\chapter_1\demo_scripts.py: Generating and plotting a chirp signal

```python
def chirp_demo():
    """
    Generating and plotting a chirp signal
    """
    import numpy as np
    import matplotlib.pyplot as plt
    from scipy.signal import chirp
    
    fs = 500 # sampling frequency in Hz
    t =np.arange(start = 0, stop = 1,step = 1/fs) #total time base from 0 to 1 second
    g = chirp(t, f0=1, t1=0.5, f1=20, phi=0, method='linear')
    plt.plot(t,g); plt.show()
```

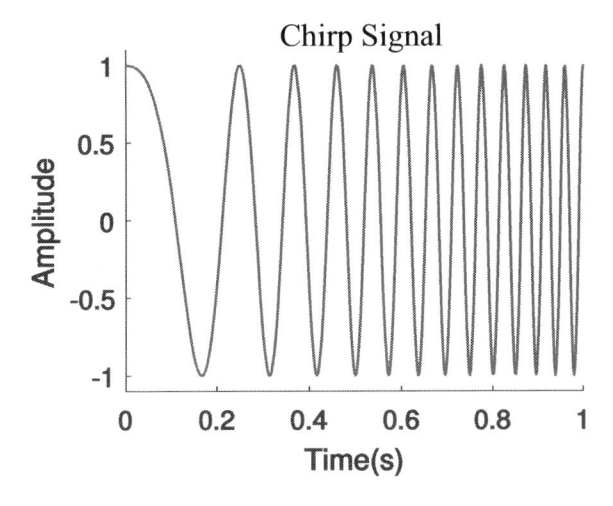

Fig. 1.5: A finite record of a chirp signal

1.2 Interpreting FFT results - complex DFT, frequency bins and FFTShift

Often, one is confronted with the problem of converting a time domain signal to frequency domain and vice-versa. Fourier transform is an excellent tool to achieve this conversion and is ubiquitously used in many applications. In signal processing, a time domain signal can be *continuous* or *discrete* and it can be *aperiodic* or *periodic*. This gives rise to four types of Fourier transforms as listed in Table 1.1.

Transform	Nature of time domain signal	Nature of frequency spectrum
Fourier Transform (FT), (a.k.a Continuous Time Fourier Transform (CTFT))	continuous, non-periodic	non-periodic, continuous
Discrete-time Fourier Transform (DTFT)	discrete, non-periodic	periodic, continuous
Fourier Series (FS)	continuous, periodic	non-periodic, discrete
Discrete Fourier Transform (DFT)	discrete, periodic	periodic, discrete

Table 1.1: Four types of Fourier transforms

From Table 1.1, we note that when the signal is discrete in one domain, it will be periodic in other domain. Similarly, if the signal is continuous in one domain, it will be aperiodic (non-periodic) in another domain. For simplicity, lets not venture into the specific equations for each of the transforms above. We will limit our discussion to DFT, that is widely available as part of software packages like SciPy, however we can approximate other transforms using DFT.

1.2.1 Real and complex DFT

For each of the listed transforms above, there exist a real version and complex version. The real version of the transform, takes in a real numbers and gives two sets of real frequency domain points - one set representing coefficients over *cosine* basis function and the other set representing the coefficient over *sine* basis function. The complex version of the transforms represent positive and negative frequencies in a single array. The complex versions are flexible, that it can process both complex valued signals and real valued signals. Figure 1.6 captures the difference between real DFT and complex DFT.

1.2.1.1 Real DFT

Consider the case of N-point *real* DFT. It takes in N samples of *real-valued* time domain waveform $x[n]$ and gives two arrays of length $N/2+1$, each set projected on cosine and sine functions respectively.

$$X_{re}[k] = \frac{2}{N} \sum_{n=0}^{N-1} x[n] cos\left(\frac{2\pi kn}{N}\right)$$
$$X_{im}[k] = -\frac{2}{N} \sum_{n=0}^{N-1} x[n] sin\left(\frac{2\pi kn}{N}\right) \quad (1.18)$$

Here, the time domain index n runs from $0 \to N$, the frequency domain index k runs from $0 \to N/2$. The real-valued time domain signal $x[n]$ can be synthesized from the real DFT pairs as

1.2 Interpreting FFT results - complex DFT, frequency bins and FFTShift

$$x[n] = \sum_{k=0}^{N/2} X_{re}[K]cos\left(\frac{2\pi kn}{N}\right) - X_{im}[K]sin\left(\frac{2\pi kn}{N}\right) \qquad (1.19)$$

Caveat: When using the synthesis equation, the values $X_{re}[0]$ and $X_{re}[N/2]$ must be divided by two. This problem is due to the fact that we restrict the analysis to real-values only. These type of problems can be avoided by using complex version of DFT.

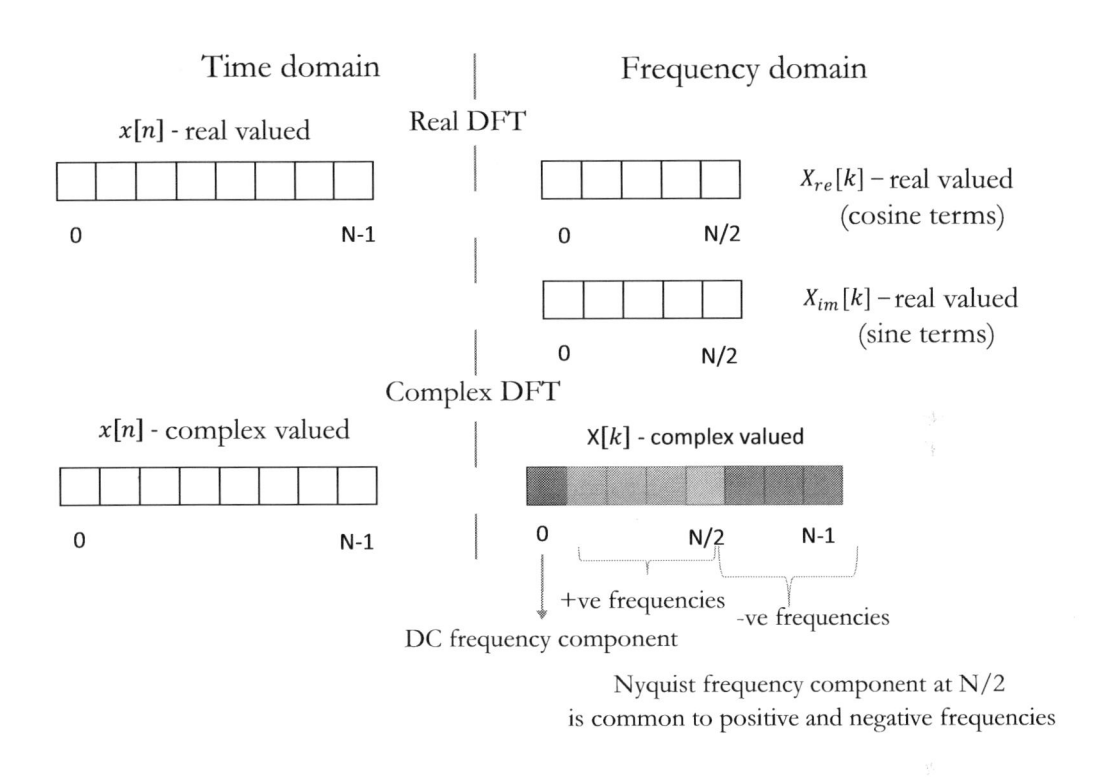

Fig. 1.6: Real and complex DFT

1.2.1.2 Complex DFT

Consider the case of N-point *complex* DFT. It takes in N samples of *complex-valued* time domain waveform $x[n]$ and produces an array $X[k]$ of length N.

$$X[k] = \frac{1}{N}\sum_{n=0}^{N-1} x[n]e^{-j2\pi kn/N} \qquad (1.20)$$

The arrays values are interpreted as follows

- $X[0]$ represents DC frequency component
- Next $N/2$ terms are positive frequency components with $X[N/2]$ being the Nyquist frequency (which is equal to half of sampling frequency)
- Next $N/2 - 1$ terms are negative frequency components (note: negative frequency components are the phasors rotating in opposite direction, they can be optionally omitted depending on the application)

The corresponding synthesis equation (reconstruct $x[n]$ from frequency domain samples $X[k]$) is

$$x[n] = \sum_{k=0}^{N-1} X[k] e^{j2\pi kn/N} \qquad (1.21)$$

From equation 1.18, we can see that the real DFT is computed by projecting the signal on cosine and sine basis functions. However, the complex DFT, in equation 1.20, projects the input signal on exponential basis functions (Euler's formula - $e^{i\theta} = cos\theta + isin\theta$ connects these two concepts).

When the input signal in the time domain is real valued, the complex DFT zero-fills the imaginary part during computation (That's its flexibility and avoids the caveat needed for real DFT). Figure 1.7 shows how to interpret the raw FFT results returned by SciPy's FFT function that computes complex DFT. The specifics will be discussed next with an example.

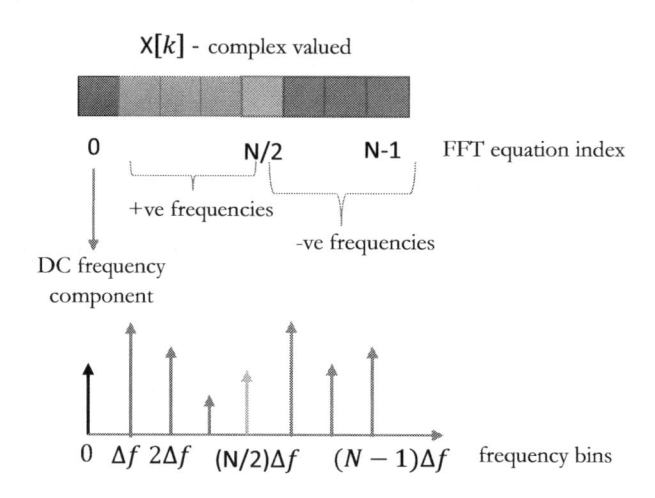

Fig. 1.7: Interpretation of frequencies in complex DFT output

1.2.2 Fast Fourier Transform (FFT)

The FFT, implemented in *Scipy.fftpack* package, is an algorithm published in 1965 by J.W.Cooley and J.W.Tuckey for efficiently calculating the DFT [1]. In its simplest implementation, parallel DFT computations are performed on the odd, even indexed samples of a record and the results are combined. If the number of samples taken for the computation is an integer power of 2, the DFT computation can be recursively performed, there by offering significant reductions in the computation time. This is called *radix* -2 FFT. The FFT length can also be odd as used in a particular FFT implementation called Prime-factor FFT algorithm [2] [3], where the FFT length factors into two co-primes.

FFT is widely available in software packages like SciPy, Matlab etc.., FFT in SciPy, implements the complex version of DFT. SciPy's FFT implementation computes the complex DFT that is very similar to the above equations, except for the scaling factor. For comparison, the SciPy's FFT implementation computes the complex DFT and its inverse as

1.2 Interpreting FFT results - complex DFT, frequency bins and FFTShift

$$X[k] = \sum_{n=0}^{N-1} x[n] e^{-j2\pi k n/N}$$

$$x[n] = \frac{1}{N} \sum_{k=0}^{N-1} X[k] e^{j2\pi k n/N} \qquad (1.22)$$

The SciPy functions that implement the above equations are FFT and IFFT respectively. The functions can be invoked as follows

```
from scipy.fftpack import fft, ifft
X = fft(x,N) #compute X[k]
x = ifft(X,N) #compute x[n]
```

1.2.3 Interpreting the FFT results

Let's assume that the $x[n]$ is a time domain cosine signal of frequency $f_c = 10\,Hz$. For representing it in the computer memory, the signal is sampled at a frequency $f_s = 32 * fc$ (Figure 1.8).

```
from scipy.fftpack import fft, ifft
import numpy as np
import matplotlib.pyplot as plt
np.set_printoptions(formatter={"float_kind": lambda x: "%g" % x})

fc=10 # frequency of the carrier
fs=32*fc # sampling frequency with oversampling factor=32
t=np.arange(start = 0,stop = 2,step = 1/fs) # 2 seconds duration
x=np.cos(2*np.pi*fc*t) # time domain signal (real number)

fig, (ax1, ax2, ax3) = plt.subplots(nrows=3, ncols=1)
ax1.plot(t,x) #plot the signal
ax1.set_title('$x[n]= cos(2 \pi 10 t)$')
ax1.set_xlabel('$t=nT_s$')
ax1.set_ylabel('$x[n]$')
```

Let's apply a $N = 2^8 = 256$ point FFT on the discrete-time signal $x[n]$.

```
N=256 # FFT size
X = fft(x,N) # N-point complex DFT, output contains DC at index 0
# Nyquist frequency at N/2 th index positive frequencies from
# index 2 to N/2-1 and negative frequencies from index N/2 to N-1
```

Note: The FFT length should be sufficient to cover the entire length of the input signal. If N is less than the length of the input signal, the input signal will be truncated when computing the FFT. The cosine wave $x[n]$ is of 2 seconds duration and it will have 640 points (a $10Hz$ frequency wave sampled at 32 times oversampling factor will have $2 \times 32 \times 10 = 640$ samples in 2 seconds of the record). Since the input signal $x[n]$ is periodic, we can safely use $N = 256$ point FFT, anyways the FFT will extend the signal when computing the FFT.

The DC component of the FFT decomposition is present at index 0.

$$x[n]=\cos(2\pi 10 t)$$

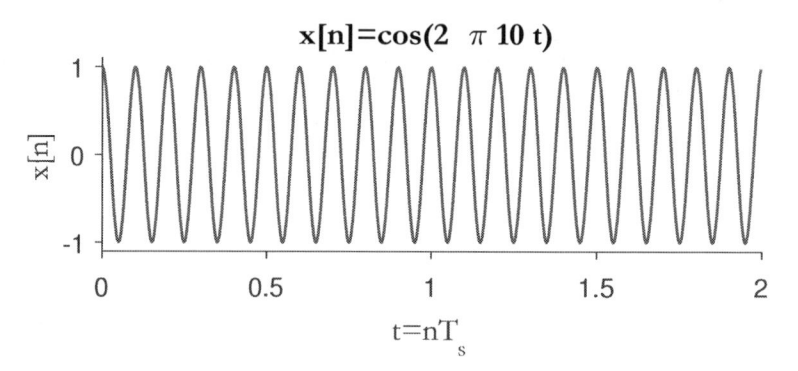

Fig. 1.8: A 2 seconds record of 10 Hz cosine wave

```
>>X[0]
(-1.3858304e-14) #approximately zero
```

Note that the index for the raw FFT are integers from $0 \to N-1$. We need to process it to convert these integers to *frequencies*. That is where the *sampling* frequency counts. Each point/bin in the FFT output array is spaced by the frequency resolution Δf, that is calculated as

$$\Delta f = \frac{f_s}{N} \tag{1.23}$$

where, f_s is the sampling frequency and N is the FFT size that is considered. Thus, for our example, each point in the array is spaced by the frequency resolution

$$\Delta f = \frac{f_s}{N} = \frac{32 * f_c}{256} = \frac{320}{256} = 1.25 Hz \tag{1.24}$$

The $10Hz$ cosine signal will register a spike at the 8th sample (10/1.25=8) - located at index 8 in Figure 1.9.

```
>>abs(X[7:10])
[0.   128.    0.]
```

Therefore, using the frequency resolution, the entire frequency axis can be computed as

```
# calculate frequency bins with FFT
df=fs/N # frequency resolution
sampleIndex = np.arange(start = 0,stop = N) # raw index for FFT plot
f=sampleIndex*df # x-axis index converted to frequencies
```

Now, plot the absolute value of the FFT against frequencies - the resulting plot is shown in the Figure 1.9.

```
ax2.stem(sampleIndex,abs(X),use_line_collection=True) # sample values on x-axis
ax2.set_title('X[k]');ax2.set_xlabel('k');ax2.set_ylabel('|X(k)|');
ax3.stem(f,abs(X),use_line_collection=True); # x-axis represent frequencies
```

1.2 Interpreting FFT results - complex DFT, frequency bins and FFTShift

```
ax3.set_title('X[f]');ax3.set_xlabel('frequencies (f)');ax3.set_ylabel('|X(f)|');
fig.show()
```

After the frequency axis is properly transformed with respect to the sampling frequency, we note that the cosine signal has registered a spike at $10Hz$. In addition to that, it has also registered a spike at $256 - 8 = 248^{th}$ sample that belongs to the negative frequency portion. Since we know the nature of the signal, we can optionally ignore the negative frequencies. The sample at the Nyquist frequency ($f_s/2$) mark the boundary between the positive and negative frequencies.

```
>>nyquistIndex=N//2 #// is for integer division
>>print(X[nyquistIndex-2:nyquistIndex+3, None]) #print array X as column
[-2.46e-14+4.27e-15j
-1.89e-14+9.72e-15j
-3.78e-14
-1.89e-14-9.72e-15j
-2.46e-14-4.27e-15j]
```

Note that the complex numbers surrounding the Nyquist index are complex conjugates and are present at positive and negative frequencies respectively.

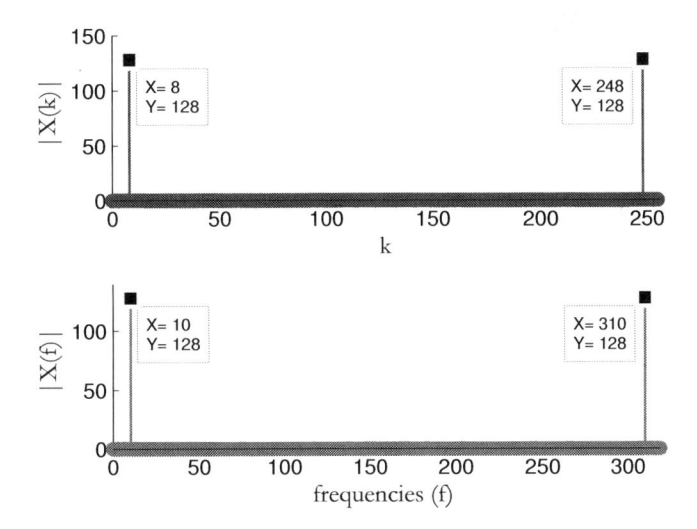

Fig. 1.9: FFT magnitude response plotted against - sample index (top) and computed frequencies (bottom)

1.2.4 FFTShift

From Figure 1.9, we see that the frequency axis starts with DC, followed by positive frequency terms which is in turn followed by the negative frequency terms. To introduce proper order in the x-axis, one can use

FFTshift function from `scipy.fftpack` package, which arranges the frequencies in order: negative frequencies → DC → positive frequencies. The fftshift function need to be carefully used when *N* is odd.

For even N, the original order returned by FFT is as follows

- $X[0]$ represents DC frequency component
- $X[1]$ to $X[N/2-1]$ terms are positive frequency components
- $X[N/2]$ is the Nyquist frequency ($F_s/2$) that is common to both positive and negative frequencies. We will consider it as part of negative frequencies to have the same equivalence with the *fftshift* function
- $X[N/2]$ to $X[N-1]$ terms are considered as negative frequency components

FFTshift shifts the DC component to the center of the spectrum. It is important to remember that the Nyquist frequency at the (N/2)th index is common to both positive and negative frequency sides. FFTshift command puts the Nyquist frequency in the negative frequency side. This is captured in the Figure 1.10. Therefore, when

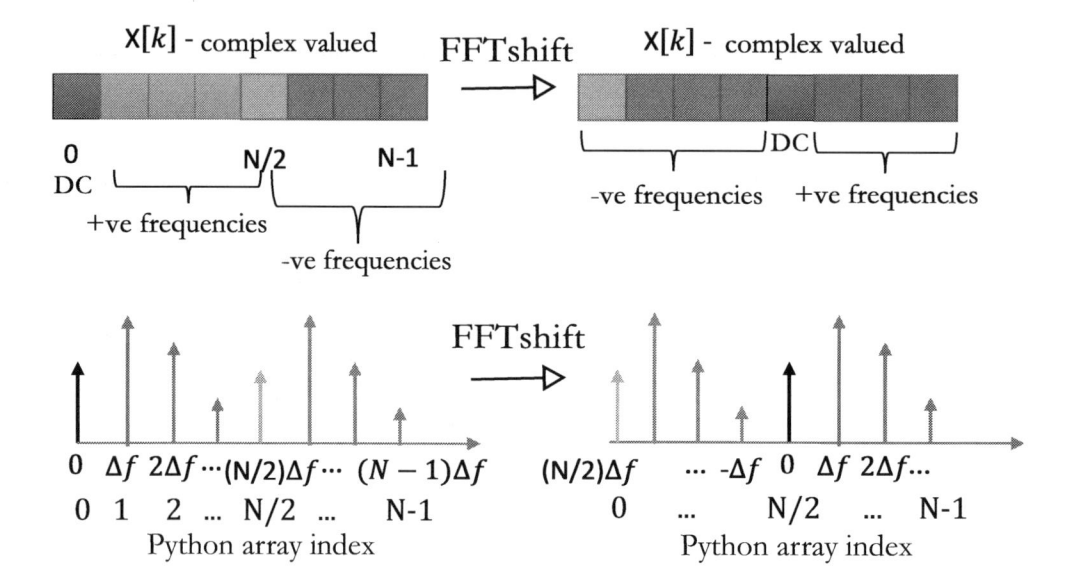

Fig. 1.10: Role of FFTShift in ordering the frequencies (assumption: N is even)

N is even, ordered frequency axis is set as

$$f = \Delta f \times i = \frac{f_s}{N} \times i, \quad where\ i = \left[-\frac{N}{2}, \cdots, -1, 0, 1, \cdots, \frac{N}{2} - 1\right] \qquad (1.25)$$

When *N* is odd, the ordered frequency axis should be set as

$$f = \Delta f \times i = \frac{f_s}{N} \times i, \quad where\ i = \left[-\frac{N+1}{2}, \cdots, -1, 0, 1, \cdots, \frac{N+1}{2} - 1\right] \qquad (1.26)$$

The following code snippet, computes the fftshift using both the manual method and using the SciPy's fftshift function. The results are plotted by superimposing them on each other. The plot in Figure 1.11 shows that both the manual method and fftshift method are in good agreement. Comparing the bottom figures in the Figure 1.9 and Figure 1.11, we see that the ordered frequency axis is more meaningful to interpret.

1.2 Interpreting FFT results - complex DFT, frequency bins and FFTShift

```
from scipy.fftpack import fftshift
#re-order the index for emulating fftshift
sampleIndex = np.arange(start = -N//2,stop = N//2) # // for integer division
X1 = X[sampleIndex] #order frequencies without using fftShift
X2 = fftshift(X) # order frequencies by using fftshift
df=fs/N # frequency resolution
f=sampleIndex*df # x-axis index converted to frequencies

#plot ordered spectrum using the two methods
fig, (ax1, ax2) = plt.subplots(nrows=2, ncols=1)#subplots creation
ax1.stem(sampleIndex,abs(X1), use_line_collection=True)# result without fftshift
ax1.stem(sampleIndex,abs(X2),'r',use_line_collection=True) #result with fftshift
ax1.set_xlabel('k');ax1.set_ylabel('|X(k)|')

ax2.stem(f,abs(X1), use_line_collection=True)
ax2.stem(f,abs(X2),'r' , use_line_collection=True)
ax2.set_xlabel('frequencies (f)'),ax2.set_ylabel('|X(f)|'); fig.show()
```

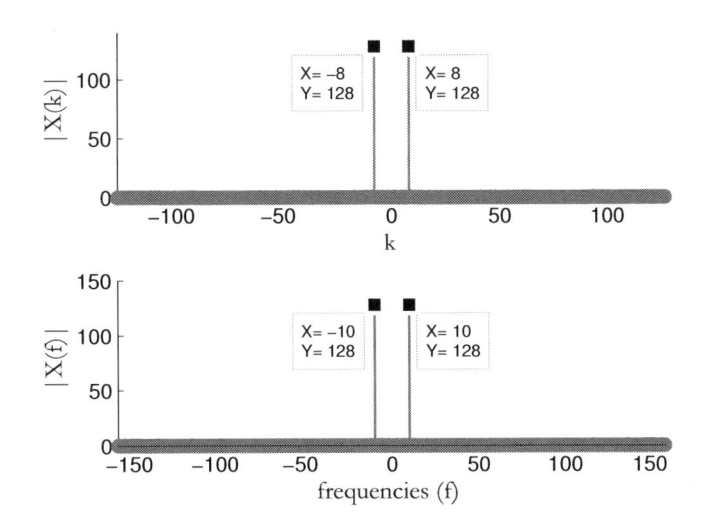

Fig. 1.11: Magnitude response of FFT result after applying FFTShift : plotted against sample index (top) and against computed frequencies (bottom)

1.2.5 IFFTShift

One can undo the effect of fftshift by employing ifftshift function. The ifftshift function restores the raw frequency order. If the FFT output is ordered using fftshift function, then one must restore the frequency components back to original order *before* taking IFFT - the Inverse Fast Fourier Transform. Following statements are equivalent.

```
X = fft(x,N) # compute X[k]
x = ifft(X,N) # compute x[n]
```

```
X = fftshift(fft(x,N)) # take FFT and rearrange frequency order
x = ifft(ifftshift(X),N) # restore raw freq order and then take IFFT
```

1.2.6 Some observations on FFTShift and IFFTShift

When N is even and for an arbitrary sequence, the `fftshift` and `ifftshift` functions will produce the same result. When they are used in tandem, it restores the original sequence.

```
>> x = np.array([0,1,2,3,4,5,6,7]) # even number of elements

>> fftshift(x)
array([4, 5, 6, 7, 0, 1, 2, 3])

>> ifftshift(x)
array([4, 5, 6, 7, 0, 1, 2, 3])

>> ifftshift(fftshift(x))
array([0, 1, 2, 3, 4, 5, 6, 7])

>> fftshift(ifftshift(x))
array([0, 1, 2, 3, 4, 5, 6, 7])
```

When N is odd and for an arbitrary sequence, the `fftshift` and `ifftshift` functions will produce different results. However, when they are used in tandem, it restores the original sequence.

```
>> x = np.array([0,1,2,3,4,5,6,7,8]) # odd number of elements

>> fftshift(x)
array([5, 6, 7, 8, 0, 1, 2, 3, 4])

>> ifftshift(x)
array([4, 5, 6, 7, 8, 0, 1, 2, 3])

>> ifftshift(fftshift(x))
array([0, 1, 2, 3, 4, 5, 6, 7, 8])

>> fftshift(ifftshift(x))
array([0, 1, 2, 3, 4, 5, 6, 7, 8])
```

1.3 Obtaining magnitude and phase information from FFT

For the discussion here, lets take an arbitrary cosine function of the form $x(t) = A\cos(2\pi f_c t + \phi)$ and proceed step by step as follows

- Represent the signal $x(t)$ in computer memory (discrete-time $x[n]$) and plot the signal in time domain
- Represent the signal in frequency domain using FFT ($X[k]$)
- Extract magnitude and phase information from the FFT result
- Reconstruct the time domain signal from the frequency domain samples

1.3.1 Discrete-time domain representation

Consider a cosine signal of amplitude $A = 0.5$, frequency $f_c = 10Hz$ and phase $\phi = \pi/6$ radians (or 30°)

$$x(t) = 0.5 \, cos\left(2\pi 10 t + \pi/6\right) \qquad (1.27)$$

In order to represent the continuous time signal $x(t)$ in computer memory (Figure 1.12), we need to sample the signal at sufficiently high rate in accordance with the Nyquist sampling theorem. I have chosen a oversampling factor of 32 so that the sampling frequency will be $f_s = 32 \times f_c$, and that gives 640 samples in a 2 seconds duration of the waveform record.

```
from scipy.fftpack import fft, ifft, fftshift, ifftshift
import numpy as np
import matplotlib.pyplot as plt

A = 0.5 # amplitude of the cosine wave
fc=10 # frequency of the cosine wave in Hz
phase=30 # desired phase shift of the cosine in degrees
fs=32*fc # sampling frequency with oversampling factor 32
t=np.arange(start = 0,stop = 2,step = 1/fs) # 2 seconds duration

phi = phase*np.pi/180; # convert phase shift in degrees in radians
x=A*np.cos(2*np.pi*fc*t+phi) # time domain signal with phase shift

fig, (ax1, ax2, ax3, ax4) = plt.subplots(nrows=4, ncols=1)
ax1.plot(t,x) # plot time domain representation
ax1.set_title(r'$x(t) = 0.5 cos (2 \pi 10 t + \pi/6)$')
ax1.set_xlabel('time (t seconds)');ax1.set_ylabel('x(t)')
```

1.3.2 Representing the signal in frequency domain using FFT

Let's represent the signal in frequency domain using the FFT function. The FFT function computes N-point complex DFT. The length of the transformation N should cover the signal of interest, otherwise we will loose some valuable information in the conversion process to frequency domain. However, we can choose a reasonable length if we know about the nature of the signal. For example, the cosine signal of our interest is periodic in nature and is of length 640 samples (for 2 seconds duration signal). We can simply use a lower

$$x(t) = 0.5 \cos(2\pi 10 t + \pi/6)$$

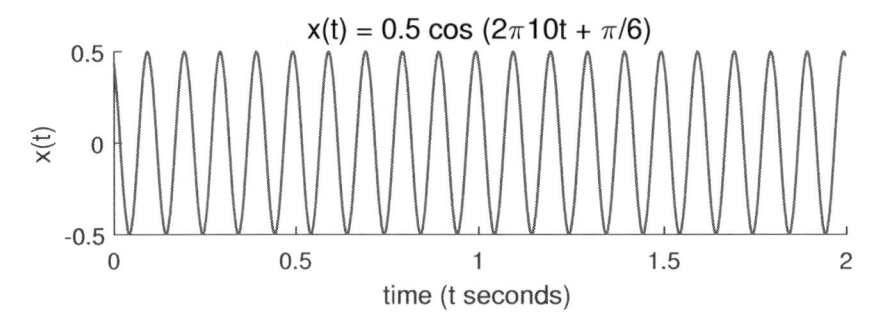

Fig. 1.12: Finite record of a cosine signal

number $N = 256$ for computing the FFT. In this case, only the first 256 time domain samples will be considered for taking FFT. However, we do not need to worry about loss of any valuable information, as the 256 samples will have sufficient number of cycles to extract the frequency of the signal.

```
N=256 # FFT size
X = 1/N*fftshift(fft(x,N)) # N-point complex DFT
```

In the code above, fftshift is used only for obtaining a nice double-sided frequency spectrum that delineates negative frequencies and positive frequencies in order. This transformation is not necessary. A scaling factor $1/N$ was used to account for the difference between the FFT implementation in Python and the text definition of complex DFT as given in equation 1.20.

1.3.2.1 Extract magnitude of frequency components (magnitude spectrum)

The FFT function computes the complex DFT and hence results in a sequence of complex numbers of form $X_{re} + jX_{im}$. The magnitude spectrum is computed as

$$|X[k]| = \sqrt{X_{re}^2 + X_{im}^2} \tag{1.28}$$

For obtaining a double-sided plot, the ordered frequency axis, obtained using fftshift, is computed based on the sampling frequency and the magnitude spectrum is plotted (Figure 1.13).

```
df=fs/N # frequency resolution
sampleIndex = np.arange(start = -N//2,stop = N//2) # // for integer division
f=sampleIndex*df # x-axis index converted to ordered frequencies
ax2.stem(f,abs(X), use_line_collection=True) # magnitudes vs frequencies
ax2.set_xlim(-30, 30)
ax2.set_title('Amplitude spectrum')
ax2.set_xlabel('f (Hz)');ax2.set_ylabel(r'$ \left| X(k) \right|$')
```

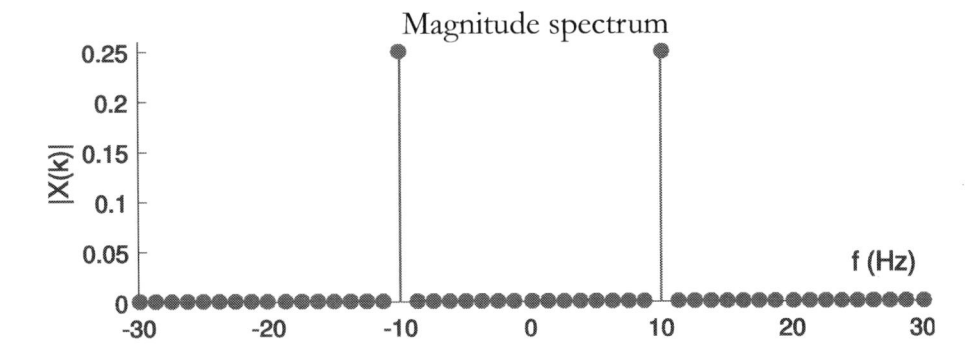

1.3 Obtaining magnitude and phase information from FFT

Fig. 1.13: Extracted magnitude information from the FFT result

1.3.2.2 Extract phase of frequency components (phase spectrum)

Extracting the correct phase spectrum is a tricky business. I will show you why it is so. The phase of the spectral components are computed as

$$\angle X[k] = tan^{-1}\left(\frac{X_{im}}{X_{re}}\right) \tag{1.29}$$

The equation 1.29 looks naive, but one should be careful when computing the inverse tangents using computers. The obvious choice for implementation seems to be the arctan function from Numpy. However, usage of arctan function will prove disastrous unless additional precautions are taken. The arctan function computes the inverse tangent over two quadrants only, i.e, it will return values only in the $[-\pi/2, \pi/2]$ interval. Therefore, the phase need to be unwrapped properly. We can simply fix this issue by computing the inverse tangent over all the four quadrants using the $arctan2(X_{img}, X_{re})$ function. Let's compute and plot the phase information using arctan2 function and see how the phase spectrum looks.

```
phase=np.arctan2(np.imag(X),np.real(X))*180/np.pi # phase information
ax3.plot(f,phase) # phase vs frequencies
```

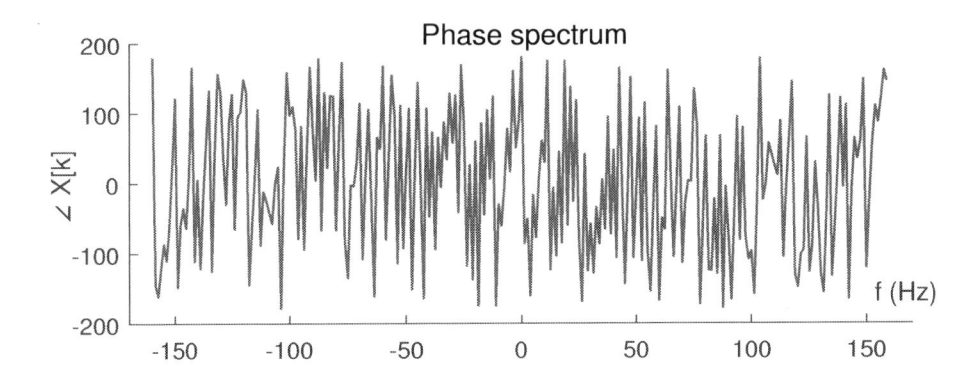

Fig. 1.14: Extracted phase information from the FFT result - phase spectrum is noisy

The phase spectrum in Figure 1.14 is completely noisy, which is unexpected. The phase spectrum is noisy due to fact that the inverse tangents are computed from the *ratio* of imaginary part to real part of the FFT result. Even a small floating rounding off error will amplify the result and manifest incorrectly as useful phase information [4]. To understand, print the first few samples from the FFT result and observe that they are not absolute zeros (they are very small numbers in the order 10^{-17}). Computing inverse tangent will result in incorrect results.

```
>>X[0:5]
[-7.2e-17  -3.6e-17-2.5e-17j  -4e-17-1.6e-17j -3.6e-17-5.6e-17j  3e-18-4.9e-17j]
>>np.arctan2(np.imag(X[0:5]),np.real(X[0:5]))
[3.14159 -2.53351 -2.8185 -2.14205 -1.50991]
```

The solution is to define a tolerance threshold and ignore all the computed phase values that are below the threshold.

```
X2=X #store the FFT results in another array
# detect noise (very small numbers (eps)) and ignore them
threshold = max(abs(X))/10000; # tolerance threshold
X2[abs(X)<threshold]=0 # maskout values below the threshold
phase=np.arctan2(np.imag(X2),np.real(X2))*180/np.pi # phase information
ax4.stem(f,phase, use_line_collection=True) # phase vs frequencies
ax4.set_xlim(-30, 30); ax4.set_title('Phase spectrum')
ax4.set_ylabel(r"$\angle$ X[k]");ax4.set_xlabel('f(Hz)')
fig.show()
```

The recomputed phase spectrum is plotted in Figure 1.15. The phase spectrum has correctly registered the $30°$ phase shift at the frequency $f = 10Hz$. The phase spectrum is anti-symmetric ($\phi = -30°$ at $f = -10Hz$), which is expected for real-valued signals.

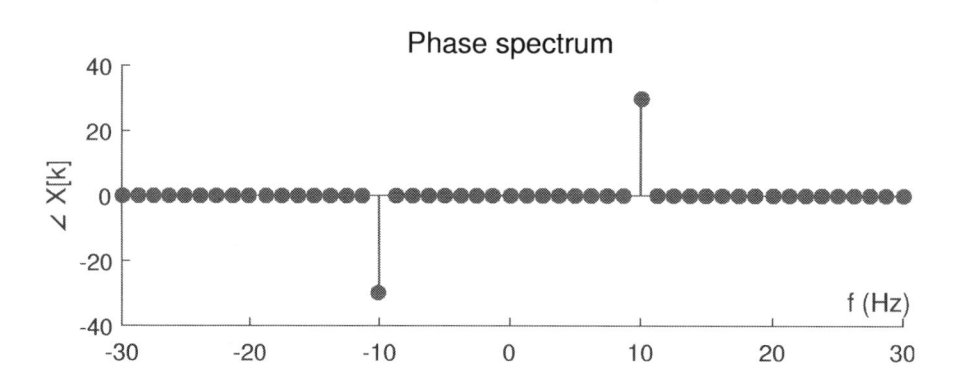

Fig. 1.15: Extracted phase information from the FFT result - recomputed phase spectrum

1.3.3 Reconstructing the time domain signal from the frequency domain samples

Reconstruction of the time domain signal from the frequency domain sample is pretty straightforward. The reconstructed signal, shown in Figure 1.16, has preserved the same initial phase shift and the frequency of the original signal. Note: The length of the reconstructed signal is only 256 samples long (≈ 0.8 seconds duration),

1.4 Power spectral density

this is because the size of FFT is considered as $N = 256$. Since the signal is periodic it is not a concern. For more complicated signals, appropriate FFT length (better to use a value that is larger than the length of the signal) need to be used.

```
x_recon = N*ifft(ifftshift(X),N) # reconstructed signal
t = np.arange(start = 0,stop = len(x_recon))/fs # recompute time index
fig2, ax5 = plt.subplots()
ax5.plot(t,np.real(x_recon)) # reconstructed signal
ax5.set_title('reconstructed signal')
ax1.set_xlabel('time (t seconds)');ax1.set_ylabel('x(t)');
fig2.show()
```

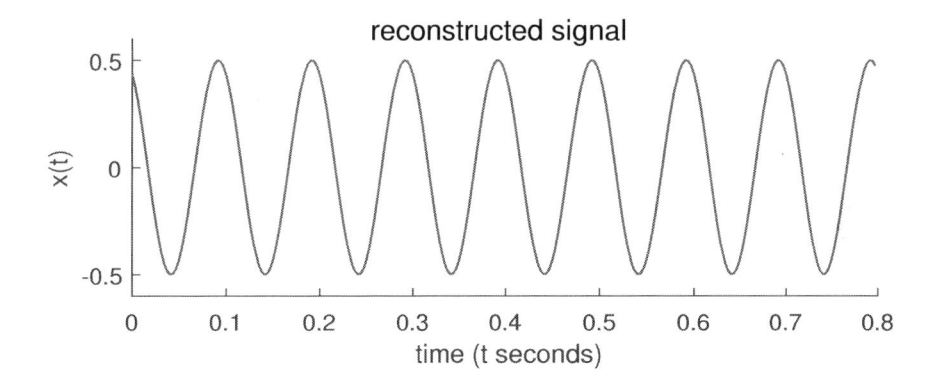

Fig. 1.16: Reconstructed time-domain signal from frequency domain samples

1.4 Power spectral density

Power spectral density (PSD) measures how the power of a signal is distributed over frequency. It is a measure of a signal's power intensity in the frequency domain. The PSD of a signal can be computed as square of the magnitude of the DFT computed in equation 1.28. However, if the signal is stochastic in nature, trying to calculate the spectral components of the signal using DFT will not be valid because, for every realization of the random process, the computed DFT will vary.

For a stochastic signal, if sufficient and accurate statistical model is available, then one can calculate its power spectral density using *Wiener-Khinchin theorem*. The theorem says that for a wide-sense stationary random process, the PSD can be calculated as the Fourier transform of the auto-correlation function of the signal.

$$S_{xx}(f) = \mathbb{F}\left[R_{xx}(\tau)\right] = \int_{-\infty}^{\infty} R_{xx}(\tau)e^{-j2\pi f\tau}d\tau \qquad (1.30)$$

where, $R_{xx}(\tau)$ is the auto-correlation function of the random process $x(t)$ given by:

$$R_{xx}(\tau) = \mathbb{E}\left(X(t)X(t-\tau)\right) = \int_{-\infty}^{\infty} x(t)x(t+\tau)dt \qquad (1.31)$$

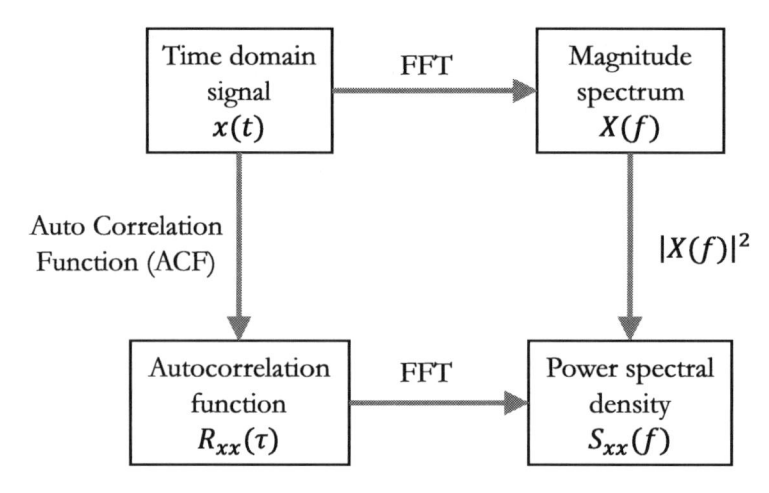

Fig. 1.17: Visualizing Weiner-Khinchin theorem

We can prove that computing PSD is equivalent to Fourier transform of the auto-correlation function of the stochastic signal $x(t)$ as follows:

$$\mathbb{F}\left[R_{xx}(\tau)\right] = \int_{-\infty}^{\infty} R_{xx}(\tau) e^{-j2\pi f \tau} d\tau \tag{1.32}$$

$$= \int_{-\infty}^{\infty} \int_{-\infty}^{\infty} x(t) x(t+\tau) e^{-j2\pi f \tau} dt d\tau$$

$$= \int_{-\infty}^{\infty} x(t) \underbrace{\int_{-\infty}^{\infty} x(t+\tau) e^{-j2\pi f \tau} d\tau}_{\mathbb{F}[x(t+\tau)] = X(f) e^{j2\pi f t}} dt$$

$$= X(f) \int_{-\infty}^{\infty} x(t) e^{j2\pi f t} dt$$

$$= X(f) X^*(f) = |X(f)|^2$$

For a stochastic signal in analysis, if sufficient statistical details are not available (which is often the practical case), one must *estimate* the power spectral density. Several methods are available for PSD estimation and the Welch method is one among them. Welch method estimates the power spectrum of a signal by dividing it into overlapping blocks, computing the *periodogram* (Fourier transform of estimate of the autocorrelation sequence) and averaging the results. The description of the Welch method and its implementation is beyond the scope of this book. In Python, the *welch* function from *scipy.signal* package can be used to estimate the Welch PSD.

The following function readily plots the Welch spectrum estimate using the SciPy's `welch` function. The function uses the recommended settings from [5], where the `welch` function is configured with Hanning window, without overlap and an averaging factor of 16. The given function is utilized in chapter 2 for plotting the PSD estimates of signals modulated using various digital modulation techniques, namely, BPSK, QPSK, MSK and GMSK (refer sections 2.10 and 2.11.3).

1.5 Power and energy of a signal

> **Program 8: DigiCommPy\essentials.py**: Estimate and plot Welch PSD for a carrier modulated signal

```
def plotWelchPSD(x,fs,fc,ax = None,color='b', label=None):
    """
    Plot PSD of a carrier modulated signal using Welch estimate
    Parameters:
        x : signal vector (numpy array) for which the PSD is plotted
        fs : sampling Frequency
        fc : center carrier frequency of the signal
        ax : Matplotlib axes object reference for plotting
        color : color character (format string) for the plot
    """
    from scipy.signal import hanning, welch
    from numpy import log10
    nx = max(x.shape)
    na = 16 # averaging factor to plot averaged welch spectrum
    w = hanning(nx//na) #// is for integer floor division
    # Welch PSD estimate with Hanning window and no overlap
    f, Pxx = welch(x,fs,window = w,noverlap=0)
    indices = (f>=fc) & (f<4*fc)   # To plot PSD from Fc to 4*Fc
    Pxx = Pxx[indices]/Pxx[indices][0] # normalized psd w.r.t Fc
    ax.plot(f[indices]-fc,10*log10(Pxx),color,label=label) #Plot in the given axes
```

1.5 Power and energy of a signal

1.5.1 Energy of a signal

In signal processing, a signal is viewed as a function of time. The term *size of a signal* is used to represent *strength of the signal*. It is crucial to know the *size* of a signal used in a certain application. For example, we may be interested to know the amount of electricity needed to power a LCD monitor as opposed to a CRT monitor. Both of these applications are different and have different tolerances. Thus the amount of electricity driving these devices will also be different.

A given signal's size can be measured in many ways. Given a mathematical function (or a signal equivalently), it seems that the area under the curve, described by the mathematical function, is a good measure of describing the size of a signal. A signal can have both positive and negative values. This may render areas that are negative. Due to this effect, it is possible that the computed values cancel each other totally or partially, rendering incorrect result. Thus the metric function of "area under the curve" is not suitable for defining the "size" of a signal. Now, we are left with two options : either 1) computation of the area under the absolute value of the function or 2) computation of the area under the square of the function. The second choice is favored due to its mathematical tractability and its similarity to Euclidean norm (Note: Euclidean norm - otherwise called *L2 norm* or *2-norm* [6] - is often considered in signal detection techniques - on the assumption that it provides a reasonable measure of distance between two points on signal space. It is computed as Euclidean distance in detection theory - see section 3.4.4).

Going by the second choice of viewing the size as the computation of the area under the square of the function, the energy of a continuous-time complex signal $x(t)$ is defined as

$$E_x = \int_{-\infty}^{\infty} |x(t)|^2 dt \tag{1.33}$$

If the signal $x(t)$ is real, the modulus operator in the above equation does not matter. This is called *energy* in signal processing terms. This is also a measure of signal strength. This definition can be applied to any signal (or a vector) irrespective of whether it possesses actual energy (a basic quantitative property as described by physics) or not. If the signal is associated with some physical energy, then the above definition gives the energy content in the signal. If the signal is an electrical signal, then the above definition gives the total energy of the signal (in Joules) dissipated over a 1 Ω resistor.

Actual Energy - the physical quantity

To know the actual energy (E) of the signal, one has to know the value of load Z the signal is driving and also the nature of the electrical signal (voltage or current). If $x(t)$ is a voltage signal, the above equation has to be scaled by a factor of $1/Z$.

$$E = \frac{E_x}{Z} = \frac{1}{Z} \int_{-\infty}^{\infty} |x(t)|^2 dt \tag{1.34}$$

If $x(t)$ is a current signal, the equation has to be scaled by Z

$$E = ZE_x = Z \int_{-\infty}^{\infty} |x(t)|^2 dt \tag{1.35}$$

where, Z is the impedance driven by the signal $x(t)$, E_x is the signal energy (signal processing term) and E is the actual energy of the signal (physical quantity) driving the load Z

Energy in discrete domain

In the discrete domain, the energy of the signal is given by

$$E_x \triangleq \|x\|^2 \triangleq \sum_{n=-\infty}^{\infty} |x[n]|^2 \tag{1.36}$$

where $\|x\|^2$ is referred as Euclidean norm. The energy is finite only if the above sum converges to a finite value. This implies that the signal is *squarely-summable*. Such a signal is called finite energy signal. If the given signal does not decay with respect to time (example: a continuous sine wave repeating its cycle indefinitely), the energy will be infinite and such a signal is *not squarely-summable*. We need another measurable quantity to circumvent this problem. This leads us to the notion of *Power*

1.5.2 Power of a signal

Power is defined as the amount of energy consumed per unit time. This quantity is useful if the energy of the signal goes to infinity or the signal is *not-squarely-summable*. For non-squarely-summable discrete-time signals, the power is calculated by taking the snapshot of the signal over a specific interval of time as follows

1. Take a snapshot of the signal over some finite time duration
2. Compute the energy of the signal E_x
3. Divide the energy by number of samples taken for computation N
4. Extend the limit of number of samples to infinity ($N \to \infty$). This gives the total power of the signal.

In discrete domain, the total power of the signal is given by

1.5 Power and energy of a signal

$$P_x = \lim_{N \to \infty} \frac{1}{2N+1} \sum_{n=-N}^{n=+N} |x[n]|^2 \quad (1.37)$$

Following equations are different forms of the same computation found in many text books. The only difference is the number of samples taken for computation. The denominator changes according to the number of samples taken for computation.

$$P_x = \lim_{N \to \infty} \frac{1}{2N} \sum_{n=-N}^{n=N-1} |x[n]|^2$$

$$P_x = \lim_{N \to \infty} \frac{1}{N} \sum_{n=0}^{n=N-1} |x[n]|^2$$

$$P_x = \lim_{N \to \infty} \frac{1}{N_1 - N_0 + 1} \sum_{n=N_0}^{n=N_1} |x[n]|^2 \quad (1.38)$$

1.5.3 Classification of signals

A signal can be classified based on its power or energy content. Signals having finite energy are energy signals. Power signals have finite and non-zero power.

Energy signal

A finite energy signal will have zero *total* power. When the energy is finite, the total power will be zero. For example, in equation 1.37, when the limit $N \to \infty$, the energy dilutes to zero over the infinite duration and hence the total power becomes zero.

Power signal

Signals whose total power is finite and non-zero. The energy of the power signal will be infinite. Example: periodic sequences like sinusoid. A sinusoidal signal has finite, non-zero power but infinite energy. A signal cannot be both an energy signal and a power signal.

Neither an energy signal nor a power signal

Signals can also be a cat on the wall - neither an energy signal nor a power signal. Consider a signal of increasing amplitude defined by $x[n] = n$. For such a signal, both the energy and power will be infinite. Thus, it cannot be classified either as an energy signal or as a power signal.

1.5.4 Computation of power of a signal - simulation and verification

The total power of a signal can be computed using the following equation. For other forms of equations for computing power, refer equation 1.38.

$$P_x = \lim_{N \to \infty} \frac{1}{N} \sum_{n=0}^{n=N-1} |x[n]|^2 \quad (1.39)$$

As a case study, a sine wave of amplitude A and frequency f_c is considered here.

$$x(t) = A\sin(2\pi f_c t) \quad (1.40)$$

When represented in frequency domain, it will look like the one on the right side plot in the Figure 1.18. This is evident from the fact that the sine wave can be mathematically represented by applying Euler's formula.

$$A\sin(2\pi f_c t) = A \frac{e^{j2\pi f_c t} - e^{-j2\pi f_c t}}{2j} \quad (1.41)$$

Taking the Fourier transform of $x(t)$ to represent it in frequency domain,

$$X(f) = F\left[A\sin(2\pi f_c t)\right] = \int_{-\infty}^{\infty} \left[\frac{e^{j2\pi f_c t} - e^{-j2\pi f_c t}}{2j}\right] e^{-j2\pi f t} dt = \frac{A}{2j}\left[\delta(f - f_c) - \delta(f + f_c)\right] \quad (1.42)$$

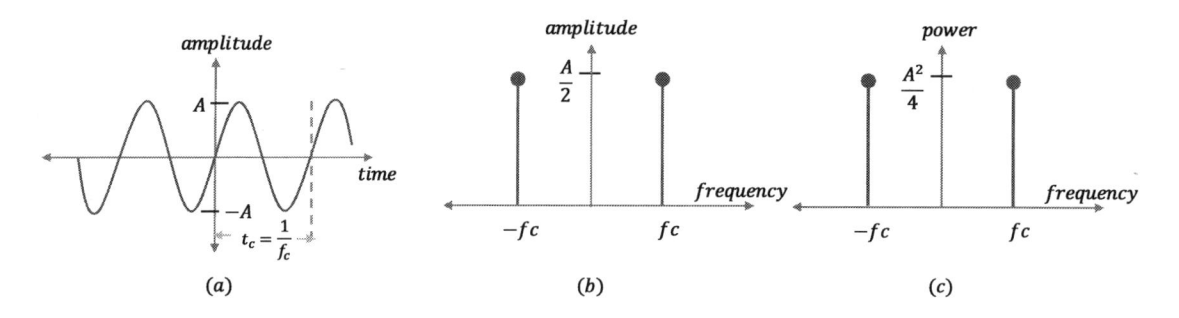

Fig. 1.18: A sinusoid represented in : (a) time domain, (b) frequency domain and (c) its power spectrum

When considering the amplitude part, the above decomposition gives two spikes of amplitude $A/2$ on either side of the frequency domain at f_c and $-f_c$ and this is shown in Figure 1.18(b). Squaring the amplitudes gives the magnitude of power of the individual frequency components. The power spectrum is shown in Figure 1.18(c).

Thus if the pure sine wave is of amplitude $A = 1\,V$ and frequency $f_c = 100\,Hz$, the power spectrum will have two spikes of value $A^2/4 = 0.25$ at $100\,Hz$ and $-100\,Hz$ frequencies. The total power will be $A^2/4 + A^2/4 = 0.25 + 0.25 = 0.5\,W$. In order to verify this through simulation, a sine wave of $100\,Hz$ frequency and amplitude $1\,V$ is taken for the experiment. The generated sine wave of 3 cycles is plotted in Figure 1.19(a).

```
import numpy as np
import matplotlib.pyplot as plt

A=1 #Amplitude of sine wave
fc=100 #Frequency of sine wave
fs=3000 # Sampling frequency - oversampled by the rate of 30
nCyl=3 # Number of cycles of the sinewave
t=np.arange(start = 0,stop = nCyl/fc,step = 1/fs) #Time base
x=-A*np.sin(2*np.pi*fc*t) # Sinusoidal function
```

1.5 Power and energy of a signal

```
fig, (ax1,ax2) = plt.subplots(nrows=1,ncols = 2)
ax1.plot(t,x)
ax1.set_title('Sinusoid of frequency $f_c=100 Hz$')
ax1.set_xlabel('Time(s)');ax1.set_ylabel('Amplitude')
fig.show()
```

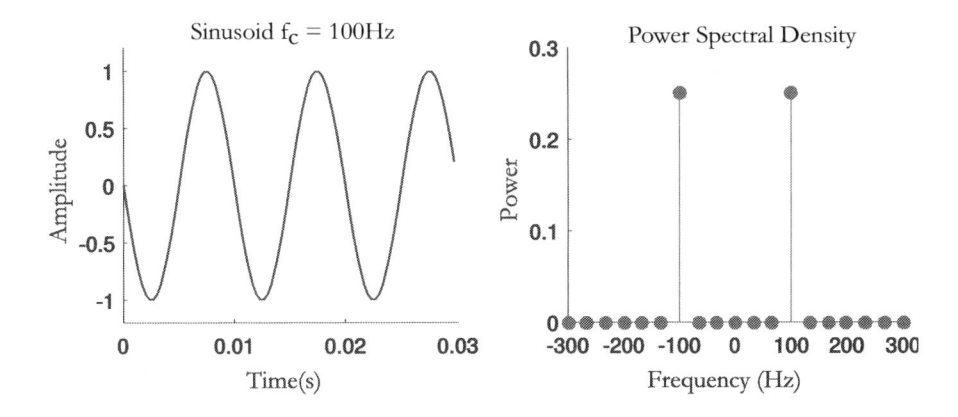

Fig. 1.19: (a) 10 cycles of the generated sine wave , (b) power spectrum

Numpy's norm function

Python's Numpy package comes with norm function. The *p-norm* in Numpy is computed as

$$NORM(v,p) = \left(\sum_{n=0}^{n=N-1} |x[v]|^p \right)^{1/p} \tag{1.43}$$

By default, the single argument norm function computes the *2-norm* as

$$NORM(v) = NORM(v,2) = \left(\sum_{n=0}^{n=N-1} |x[v]|^2 \right)^{1/2} \tag{1.44}$$

To compute the total power of the signal $x[n]$ (as given in equation 1.39), all we have to do is - compute norm(x), square it and divide by the length of the signal. It can be verified that the snippet of code given here, gives the total power as 0.5.

```
from numpy.linalg import norm
L=len(x)
P=(norm(x)**2)/L; #norm from numpy linear algo package
print('Power of the Signal from Time domain {:0.4f}'.format(P))
```

Verifying the total power in frequency domain

Here, the total power is verified by applying *Discrete Fourier Transform (DFT)* on the sinusoidal sequence. The sinusoidal sequence $x[n]$ is represented in frequency domain $X[f]$ using SciPy's FFT function. The power associated with each frequency point is computed as

$$P_x[f] = X[f]X^*[f] \tag{1.45}$$

Finally, the total power is calculated as the sum of all the points in the frequency domain representation. The following code will result in the calculated total power equaling a value of 0.5 and the power spectrum plot as given in Figure 1.19(b)

```
from scipy.fftpack import fft,fftshift
NFFT=L
X=fftshift(fft(x,NFFT))
Px=X*np.conj(X)/(L**2) #Power of each freq components
fVals=fs*np.arange(start = -NFFT/2,stop = NFFT/2)/NFFT
ax2.stem(fVals,Px,'r')
ax2.set_title('Power Spectral Density');
ax2.set_xlabel('Frequency (Hz)');ax2.set_ylabel('Power')
```

1.6 Polynomials, convolution and Toeplitz matrices

Convolution operation is ubiquitous in signal processing applications. The mathematics of convolution is strongly rooted in operation on polynomials. The intent of this text is to enhance the understanding on mathematical details of convolution.

1.6.1 Polynomial functions

Polynomial functions are expressions consisting of sum of terms, where each term includes one or more variables raised to a non-negative power and each term may be scaled by a coefficient. Addition, subtraction and multiplication of polynomials are possible.

Polynomial functions can involve one or more variables. For example, following polynomial expression is a function of variable x. It involves sum of 3 terms where each term is scaled by a coefficient.

$$f(x) = x^2 + 2x + 1 \tag{1.46}$$

Polynomial expression involving two variables x and y is given next.

$$f(x,y) = 2x^4 - 4x^2y + 3xy^2 + 8y^2 + 5 \tag{1.47}$$

1.6.2 *Representing single variable polynomial functions*

Polynomial functions involving single variable is of specific interest here. In general, a single variable (say x) polynomial is expressed in the following sum of terms form, where $a_0, a_1, a_2, ..., a_{n-1}$ are coefficients of the polynomial.

$$f(x) = a_0 + a_1 x + a_2 x^2 + a_3 x^3 + ... + a_{n-1} x^{n-1} \tag{1.48}$$

The degree or order of the polynomial function is the highest power of x with a non-zero coefficient. The above equation can be written as

$$f(x) = \sum_{i=0}^{n-1} a_i x^i \tag{1.49}$$

It can be represented by a vector of coefficients as $a = [a_0, a_1, a_2, ..., a_{n-1}]$. Polynomials can also be represented using their roots which is a product of linear terms form, as explained next.

1.6.3 *Multiplication of polynomials and linear convolution*

Mathematical operations like addition, subtraction and multiplication can be performed on polynomial functions. Addition or subtraction of polynomials is straight forward. Multiplication of polynomials is of specific interest in the context of subject discussed here. Let's consider two polynomials represented by the coefficient vectors $a = [a_0, a_1, a_2, ..., a_n]$ and $b = [b_0, b_1, b_2, ..., b_n]$.

$$p(x) = a_0 + a_1 x + a_2 x^2 + a_3 x^3 + ... + a_n x^n \tag{1.50}$$

$$q(x) = b_0 + b_1 x + b_2 x^2 + a_3 x^3 + ... + b_m x^m \tag{1.51}$$

The product vector is expressed as

$$p(x).q(x) = a_0 b_0 + (a_1 b_0 + b_1 a_0) x + ... + a_n b_m x^{n+m} \tag{1.52}$$

or equivalently,

$$p(x).q(x) = (p*q)(x) = a.b = [c_0, c_1, c_2, ..., c_{n+m}] \tag{1.53}$$

where,

$$c_k = \sum_{i,j: i+j=k} a_i b_j \quad k = 0, 1, ..., n+m \tag{1.54}$$

Since the subscripts obey the equality $i+j = k$, changing the subscript j to $k-i$ gives

$$c_k = \sum_{i=-\infty}^{\infty} a_i b_{k-i} \quad k = 0, 1, ..., n+m \tag{1.55}$$

which, when written in terms of array index, provides the most widely used form seen in signal processing text books.

$$c[k] = \sum_{i=-\infty}^{\infty} a[i] b[k-i] \quad k = 0, 1, ..., n+m \tag{1.56}$$

This operation is referred as *linear convolution*, denoted by the symbol $*$. It is very closely related to other operations on vectors like cross-correlation, auto-correlation and moving average computation. Thus, when we are computing convolution, we are actually multiplying two polynomials. Note, that if the polynomials have N and M terms, their multiplication produces $N + M - 1$ terms.

1.6.4 Toeplitz matrix and convolution

Convolution operation of two sequences can also be viewed as multiplying two matrices. Given a *linear time invariant* (LTI) system with impulse response $h[n]$ and an input sequence $x[n]$, the output of the system $y[n]$ is obtained by *linearly* convolving the input sequence and impulse response.

$$y[k] = h[n] * x[n] = \sum_{i=-\infty}^{\infty} x[i]h[k-i] \qquad (1.57)$$

where, the sequence $x[n]$ is of length N and $h[n]$ is of length M. Assume that the sequence $h[n]$ is of length 4 given by $h[n] = [h_0, h_1, h_2, h_3]$ and the sequence $x[n]$ is of length 3 given by $x[n] = [x_0, x_1, x_2]$. The convolution $h[n] * x[n]$ is given by

$$y[k] = h[n] * x[n] = \sum_{i=-\infty}^{\infty} x[i]h[k-i] \qquad k=0,1,\ldots,5 \qquad (1.58)$$

Computing each sample in the convolution,

$$y[0] = \sum_{i=-\infty}^{\infty} x[i]h[-i] = x[0]h[0] + 0 + 0$$

$$y[1] = \sum_{i=-\infty}^{\infty} x[i]h[1-i] = x[0]h[1] + x[1]h[0] + 0$$

$$y[2] = \sum_{i=-\infty}^{\infty} x[i]h[2-i] = x[0]h[2] + x[1]h[1] + x[2]h[0]$$

$$y[3] = \sum_{i=-\infty}^{\infty} x[i]h[3-i] = x[0]h[3] + x[1]h[2] + x[2]h[1]$$

$$y[4] = \sum_{i=-\infty}^{\infty} x[i]h[4-i] = x[1]h[3] + x[2]h[1] + 0$$

$$y[5] = \sum_{i=-\infty}^{\infty} x[i]h[5-i] = x[2]h[3] + 0 + 0 \qquad (1.59)$$

Note the above result. See how the series $x[n]$ and $h[n]$ multiply with each other from the opposite directions. This gives the reason on why we have to reverse one of the sequences and shift one step a time to do the convolution operation.

Thus, graphically, in convolution, one of the sequences (say $h[n]$) is reversed. It is delayed to the extreme left where there are no overlaps between the two sequences. Now, the sample offset of $h[n]$ is increased 1 step at a time. At each step, the overlapping portions of $h[n]$ and $x[n]$ are multiplied and summed. This process is repeated until the sequence $h[n]$ is slid to the extreme right where no more overlaps between $h[n]$ and $x[n]$ are possible.

Representing the equation 1.59 in matrix form,

$$\begin{bmatrix} y[0] \\ y[1] \\ y[2] \\ y[3] \\ y[4] \\ y[5] \end{bmatrix} = \begin{bmatrix} h[0] & 0 & 0 \\ h[1] & h[0] & 0 \\ h[2] & h[1] & h[0] \\ h[3] & h[2] & h[1] \\ 0 & h[3] & h[2] \\ 0 & 0 & h[3] \end{bmatrix} \begin{bmatrix} x[0] \\ x[1] \\ x[2] \end{bmatrix} \qquad (1.60)$$

1.7 Methods to compute convolution

When the sequences $h[n]$ and $x[n]$ are represented as matrices, the convolution operation can be equivalently represented as

$$y = h*x = x*h = \begin{bmatrix} h_0 & 0 & 0 \\ h_1 & h_0 & 0 \\ h_2 & h_1 & h_0 \\ h_3 & h_2 & h_1 \\ 0 & h_3 & h_2 \\ 0 & 0 & h_3 \end{bmatrix} \begin{bmatrix} x_0 \\ x_1 \\ x_2 \end{bmatrix} \quad (1.61)$$

The matrix representing the incremental delays of $h[n]$ used in the above equation is a special form of matrix called *Toeplitz matrix*. Toeplitz matrices have constant entries along their diagonals. Toeplitz matrices are used to model systems that possess shift invariant properties. The property of shift invariance is evident from the matrix structure itself. Since we are modeling a LTI system [7], Toeplitz matrices are our natural choice. On a side note, a special form of Toeplitz matrix called *circulant matrix* is used in applications involving *circular convolution* and *Discrete Fourier Transform (DFT)* [8].

Representing the construction of Toeplitz matrix, in equation 1.61, as a function $T(h)$,

$$T(h) = \begin{bmatrix} h_0 & 0 & 0 \\ h_1 & h_0 & 0 \\ h_2 & h_1 & h_0 \\ h_3 & h_2 & h_1 \\ 0 & h_3 & h_2 \\ 0 & 0 & h_3 \end{bmatrix} \quad (1.62)$$

the convolution of h and x is simply a matrix multiplication of Toeplitz matrix $T(h)$ and the vector representation of x denoted as X

$$y = h*x = x*h = \begin{bmatrix} h_0 & 0 & 0 \\ h_1 & h_0 & 0 \\ h_2 & h_1 & h_0 \\ h_3 & h_2 & h_1 \\ 0 & h_3 & h_2 \\ 0 & 0 & h_3 \end{bmatrix} \begin{bmatrix} x_0 \\ x_1 \\ x_2 \end{bmatrix} = T(h).X \quad (1.63)$$

One can quickly vectorize the convolution operation in Python by using the *Toeplitz* function from SciPy linear algebra package [9] as shown here.

```
from scipy.linalg import toeplitz
y=toeplitz(np.array([h0,h1,h2,h3,0,0]),np.array([h0,0,0]))@x.transpose()
#@ symbol means matrix multiplication
```

1.7 Methods to compute convolution

Mathematical details of convolution, its relationship to polynomial multiplication and the application of Toeplitz matrices in computing linear convolution are discussed in the previous section (1.6) of this chapter. A short survey of different techniques to compute discrete linear convolution is given here.

Given a linear time invariant (LTI) system with impulse response $h[n]$ and an input sequence $x[n]$, the output of the system $y[n]$ is obtained by convolving the input sequence and impulse response.

$$y[k] = h[n] * x[n] = \sum_{i=-\infty}^{\infty} x[i]h[k-i] \qquad (1.64)$$

where, the sequence $x[n]$ is of length N and $h[n]$ is of length M. SciPy package contains an in-built function called convolve for computing convolution. The following methods can also be substituted for computing convolution between any two given sequences.

1.7.1 Method 1: Brute-force method

As shown in section 1.6.3, the convolution equation 1.64 can simply be interpreted as polynomial multiplication and hence it can be implemented using nested for-loops. However this method consumes the highest computational time of all the methods given here. Typically, the computational complexity is $O(n^2)$ time.

Program 9: DigiCommPy\essentials.py: Brute force method to compute convolution

```python
def conv_brute_force(x,h):
    """
    Brute force method to compute convolution
    Parameters:
        x, h : numpy vectors
    Returns:
        y : convolution of x and h
    """
    N=len(x)
    M=len(h)
    y = np.zeros(N+M-1) #array filled with zeros
    for i in np.arange(0,N):
        for j in np.arange(0,M):
            y[i+j] = y[i+j] + x[i] * h[j]
    return y
```

1.7.2 Method 2: Using Toeplitz matrix

When the sequences $h[n]$ and $x[n]$ are represented as matrices, the convolution operation can be equivalently represented as

$$y = h * x = x * h = Toeplitz(h).X = \mathbf{T}(h).\mathbf{X} \qquad (1.65)$$

When the convolution of two sequences of lengths N and p is computed, the Toeplitz matrix $\mathbf{T}(h)$ is of size $(N+p-1) \times (p)$.

The following generic function constructs a Toeplitz matrix of size $(N+p-1) \times (p)$ from a given sequence of length N. Refer section 1.6.4 for the mathematical details of Toeplitz matrix and its relationship to convolution operation.

1.7 Methods to compute convolution

Program 10: DigiCommPy\essentials.py: Construct Toeplitz matrix of size (N+p-1)xp

```python
def convMatrix(h,p):
    """
    Construct the convolution matrix of size (N+p-1)x p from the input matrix h of
    size N.
    Parameters:
        h : numpy vector of length N
        p : scalar value
    Returns:
        H : convolution matrix of size (N+p-1)xp
    """
    col=np.hstack((h,np.zeros(p-1)))
    row=np.hstack((h[0],np.zeros(p-1)))
    
    from scipy.linalg import toeplitz
    H=toeplitz(col,row)
    return H
```

A generalized function called my_convolve, given here, finds the convolution of two sequences of arbitrary lengths. It makes uses of the function convMatrix that was just described above.

Program 11: DigiCommPy\essentials.py: Computing convolution of two sequences

```python
def my_convolve(h,x):
    """
    Convolve two sequences h and x of arbitrary lengths: y=h*x
    Parameters:
        h,x : numpy vectors
    Returns:
        y : convolution of h and x
    """
    H=convMatrix(h,len(x)) #see convMatrix function
    y=H @ x.transpose() # equivalent to np.convolve(h,x) function
    return y
```

1.7.3 Method 3: Using FFT to compute convolution

Computation of convolution using FFT (Fast Fourier Transform) has the advantage of reduced computational complexity when the length of the input vectors are large. To compute convolution, take FFT of the two sequences, **x** of length N and **h** of length M with FFT length set to convolution output length satisfying $2^L \geq |N+M-1|$, multiply the results and convert back to time-domain using IFFT (Inverse Fast Fourier Transform). Note that FFT manifests as *circular convolution* in time domain. But, we are attempting to compute linear convolution using circular convolution (or FFT) with zero-padding either one of the input sequence. This causes inefficiency when compared to circular convolution. Nevertheless, this method still provides $O\left(\frac{N}{log_2 N}\right)$ savings over brute-force method.

$$y[n] = IFFT[FFT_L(\mathbf{x}) * FFT_L(\mathbf{h})] \qquad 2^L \geq |N+M-1| \qquad (1.66)$$

The following algorithm, which ignores additional zeros in the output terms, is sufficient.

$$y[n] = IFFT[FFT_L(\mathbf{x}) * FFT_L(\mathbf{h})] \qquad L = N + M - 1 \qquad (1.67)$$

Python code snippet for implementing the above algorithm is as follows.

```
from scipy.fftpack import fft,ifft
y=ifft(fft(x,L)*(fft(h,L))) #Convolution using FFT and IFFT
```

Test and comparison

Let's test the convolution methods, especially the method 2 (convolution using Toeplitz matrix transformation) and method 3 (convolution using FFT) by comparing them against Numpy's standard convolve function.

Program 12: DigiCommPy\chapter_1\demo_scripts.py: Comparing different methods for convolution

```
def compare_convolutions():
    """
    Comparing different methods for computing convolution
    """
    import numpy as np
    from scipy.fftpack import fft,ifft
    from DigiCommPy.essentials import my_convolve #import our function from
    ↪ essentials.py

    x = np.random.normal(size = 7) + 1j*np.random.normal(size = 7) #normal random
    ↪ complex vectors
    h = np.random.normal(size = 3) + 1j*np.random.normal(size = 3) #normal random
    ↪ complex vectors
    L = len(x) + len(h) - 1 #length of convolution output

    y1=my_convolve(h,x) #Convolution Using Toeplitz matrix
    y2=ifft(fft(x,L)*(fft(h,L))).T #Convolution using FFT
    y3=np.convolve(h,x) #Numpy's standard function
    print(f' y1 : {y1} \n y2 : {y2} \n y3 : {y3} \n')
```

On comparing method 2 (output $y1$) and method 3 (output $y2$) with SciPy's standard convolution function (output $y3$), it is found that all three methods yield identical results. Results obtained on a sample run are given here.

$$x = \begin{bmatrix} 0.5404 + 0.2147i \\ -0.0915 + 2.0108i \\ -0.7603 + 0.0256i \\ -0.6936 + 0.3083i \\ 1.2815 - 0.9382i \\ -0.8097 + 1.6742i \\ -1.2368 + 0.1250i \end{bmatrix} ; \; h = \begin{bmatrix} 0.5301 + 0.3891i \\ -0.9521 - 1.1560i \\ 0.8540 + 0.0397i \end{bmatrix} ; \; y1 = y2 = y3 = \begin{bmatrix} 0.2029 + 0.3241i \\ -1.0973 + 0.2012i \\ 2.4516 - 1.8860i \\ 0.1076 + 2.4617i \\ 1.4109 + 0.5012i \\ -3.9900 + 0.2200i \\ 3.1337 - 1.8233i \\ 0.5639 + 2.7084i \\ -1.0613 + 0.0576i \end{bmatrix} \qquad (1.68)$$

1.7.4 Miscellaneous methods

If the input sequence is of infinite length or very large as in many real time applications, block processing methods like *Overlap-Add* and *Overlap-Save* can be used to compute convolution in a faster and efficient way. There exist standard algorithms for computing convolution that greatly reduce the computational complexity. Some of them are given here for reference. Refer [10] for more details.

- Cook-Toom algorithm
- Modified Cook-Toom algorithm
- Winograd algorithm
- Modified Winograd algorithm
- Iterated convolution

1.8 Analytic signal and its applications

1.8.1 Analytic signal and Fourier transform

Fourier transform of a real-valued signal is complex-symmetric. It implies that the content at negative frequencies are redundant with respect to the positive frequencies. In their works, Gabor [11] and Ville [12], aimed to create an *analytic signal* by removing redundant negative frequency content resulting from the Fourier transform. The analytic signal is complex-valued but its spectrum is one-sided (only positive frequencies) that preserves the spectral content of the original real-valued signal. Using an analytic signal instead of the original real-valued signal, has proven to be useful in many signal processing applications. For example, in spectral analysis, use of analytic signal in-lieu of the original real-valued signal mitigates estimation biases and eliminates cross-term artifacts due to negative and positive frequency components [13].

1.8.1.1 Continuous-time analytic signal

Let $x(t)$ be a real-valued non-band-limited finite energy signal, for which we wish to construct a corresponding analytic signal $z(t)$. The *continuous time fourier transform* (CTFT) of $x(t)$ is given by

$$X(f) = \int_{-\infty}^{\infty} x(t) e^{-j2\pi ft} dt \qquad (1.69)$$

Let's say the magnitude spectrum of $X(f)$ is as shown in Figure 1.20(a). We note that the signal $x(t)$ is a real-valued and its magnitude spectrum $|X(f)|$ is symmetric and extends infinitely in the frequency domain.

As mentioned before, an analytic signal can be formed by suppressing the negative frequency contents of the Fourier transform of the real-valued signal. That is, in frequency domain, the spectral content $Z(f)$ of the analytic signal $z(t)$ is given by

$$Z(f) = \begin{cases} X(0) & for\ f = 0 \\ 2X(f) & for\ f > 0 \\ 0 & for\ f < 0 \end{cases} \qquad (1.70)$$

The corresponding spectrum of the resulting analytic signal is shown in Figure 1.20(b).

Since the spectrum of the analytic signal is one-sided, the analytic signal will be complex valued in the time domain, hence the analytic signal can be represented in terms of real and imaginary components as $z(t) = z_r(t) + jz_i(t)$. Since the spectral content is preserved in an analytic signal, it turns out that the real part

|X(f)|

|Z(f)|

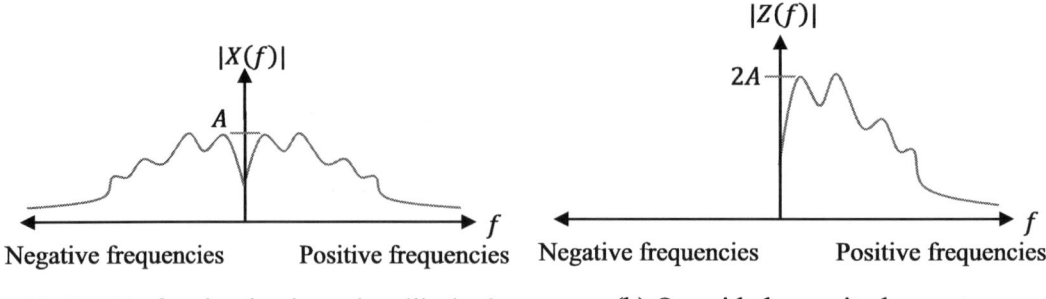

(a) CTFT of real-valued non-bandlimited finite energy signal x(t)

(b) One sided magnitude spectrum of the corresponding analytic signal z(t)

Fig. 1.20: (a) Spectrum of continuous signal x(t) and (b) spectrum of analytic signal z(t)

of the analytic signal in time domain is essentially the original real-valued signal itself ($z_r(t) = x(t)$). Then, what takes the place of the imaginary part ? What accompanies $x(t)$, that occupies the imaginary part in the resulting analytic signal ? Summarizing the question as equation,

$$z(t) = z_r(t) + jz_i(t) \tag{1.71}$$

$$z_r(t) = x(t) \quad z_i(t) = ?? \tag{1.72}$$

It is interesting to note that Hilbert transform [14] can be used to find a companion function (imaginary part in the equation 1.72) to a real-valued signal such that the real signal can be analytically extended from the real axis to the upper half of the complex plane. Denoting Hilbert transform as $HT\{\cdot\}$, the analytic signal is given by

$$z(t) = z_r(t) + jz_i(t) = x(t) + jHT\{x(t)\} \tag{1.73}$$

One important property of an analytic signal is that its real and imaginary components are orthogonal.

$$\int_{-\infty}^{\infty} z_i(t)z_r(t) = 0 \tag{1.74}$$

From these discussions, we can see that an analytic signal $z(t)$ for a real-valued signal $x(t)$, can be constructed using two approaches.

- **Frequency domain approach**: The one-sided spectrum of $z(t)$ is formed from the two-sided spectrum of the real-valued signal $x(t)$ by applying equation 1.70.
- **Time domain approach**: Using Hilbert transform approach given in equation 1.73.

1.8.1.2 Discrete-time analytic signal

Consider a continuous real-valued signal $x(t)$, that gets sampled at interval T seconds and results in N real-valued discrete samples $x[n]$, i.e, $x[n] = x(nT)$. The spectrum of the continuous signal is shown in Figure 1.21(a). The spectrum of $x[n]$ that results from the process of periodic sampling is given in Figure 1.21(b). The spectrum of discrete-time signal $x[n]$ can be obtained by *discrete-time fourier transform* (DTFT).

$$X(f) = T \sum_{n=0}^{N-1} x[n]e^{-j2\pi f nT} \tag{1.75}$$

1.8 Analytic signal and its applications

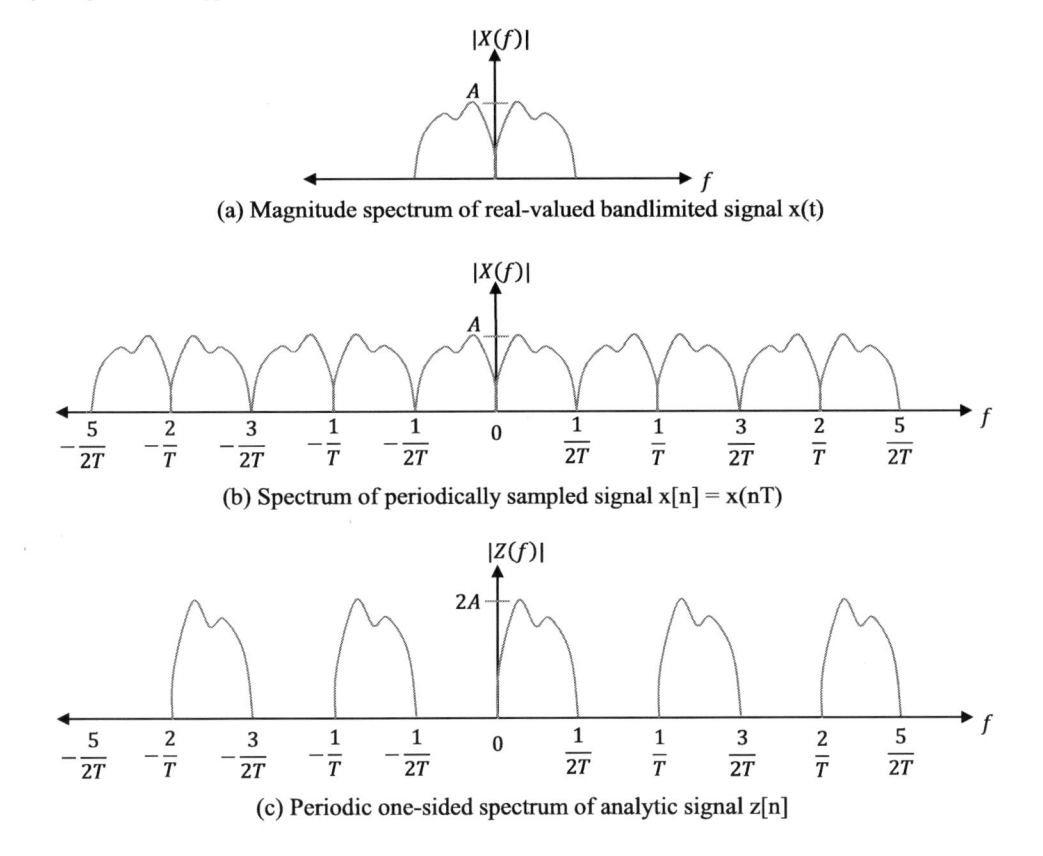

Fig. 1.21: (a) Spectrum of continuous signal x(t), (b) Spectrum of x[n] resulted due to periodic sampling and (c) Periodic one-sided spectrum of analytical signal z[n].

At this point, we would like to construct a discrete-time analytic signal $z[n]$ from the real-valued sampled signal $x[n]$. We wish the analytic signal is complex valued $z[n] = z_r[n] + jz_i[n]$ and should satisfy the following two desired properties

- The real part of the analytic signal should be same as the original real-valued signal.

$$z_r[n] = x[n] \tag{1.76}$$

- The real and imaginary part of the analytic signal should satisfy the property of orthogonality.

$$\sum_{n=0}^{N-1} z_r[n]z_i[n] = 0 \tag{1.77}$$

In frequency domain approach for the continuous-time case, we saw that an analytic signal is constructed by suppressing the negative frequency components from the spectrum of the real signal. We cannot do this for our periodically sampled signal $x[n]$. Periodic mirroring nature of the spectrum prevents one from suppressing the negative components. If we do so, it will suppress the entire spectrum. One solution to this problem is to set the negative half of each spectral period to zero. The resulting spectrum of the analytic signal is shown in Figure 1.21(c).

Given a record of samples $x[n]$ of even length N, the procedure to construct the analytic signal $z[n]$ is as follows [15].

- Compute the N-point DTFT of $x[n]$ using FFT
- N-point periodic one-sided analytic signal is computed by the following transform

$$Z[m] = \begin{cases} X[0] & for\ m = 0 \\ 2X[m] & for\ 1 \leq m \leq \frac{N}{2} - 1 \\ X[\frac{N}{2}] & for\ m = \frac{N}{2} \\ 0 & for\ \frac{N}{2} + 1 \leq m \leq N - 1 \end{cases} \qquad (1.78)$$

- Finally, the analytic signal $z[n]$ is obtained by taking the inverse DTFT of $Z[m]$

$$z[n] = \frac{1}{NT} \sum_{m=0}^{N-1} z[m]\, exp\left(j2\pi mn/N\right) \qquad (1.79)$$

This method satisfies both the desired properties listed in equations 1.76 and 1.77. The given procedure can be coded in Python using the fft function from scipy.fftpack package. Given a record of N real-valued samples $x[n]$, the corresponding analytic signal $z[n]$ can be constructed as given next.

Program 13: DigiCommPy\essentials.py: Generating an analytic signal for a given real-valued signal

```
def analytic_signal(x):
    """
    Generate analytic signal using frequency domain approach
    Parameters:
        x : Signal data. Must be real
    Returns:
        z : Analytic signal of x
    """
    from scipy.fftpack import fft,ifft
    N = len(x)
    X = fft(x,N)
    Z = np.hstack((X[0], 2*X[1:N//2], X[N//2], np.zeros(N//2-1)))
    z = ifft(Z,N)
    return z
```

To test this function, we create a 5 seconds record of a real-valued sine signal and pass it as an argument to the function. The resulting analytic signal is constructed and its orthogonal components are plotted in Figure 1.22. From this plot, we can see that the real part of the analytic signal is identical to the original signal and the imaginary part of the analytic signal is $-90°$ phase shifted version of the original signal. We note that the analytic signal's imaginary part is Hilbert transform of its real part.

Program 14: DigiCommPy\chapter_1\demo_scripts.py: Investigate components of an analytic signal

```
def analytic_signal_demo():
    """
    Investigate components of an analytic signal
    """
    import numpy as np
    import matplotlib.pyplot as plt
    from DigiCommPy.essentials import analytic_signal #import our function from
    ↪    essentials.py
```

1.8 Analytic signal and its applications

```python
t=np.arange(start=0,stop=0.5,step=0.001); # time base
x = np.sin(2*np.pi*10*t) # real-valued f = 10 Hz

fig, (ax1, ax2) = plt.subplots(nrows=2, ncols=1)
ax1.plot(t,x) # plot the original signal
ax1.set_title('x[n] - real-valued signal')
ax1.set_xlabel('n');
ax1.set_ylabel('x[n]')

z = analytic_signal(x) # construct analytic signal

ax2.plot(t, np.real(z), 'k',label='Real(z[n])')
ax2.plot(t, np.imag(z), 'r',label='Imag(z[n])')
ax2.set_title('Components of Analytic signal');
ax2.set_xlabel('n');
ax2.set_ylabel(r'$z_r[n]$ and $z_i[n]$')
ax2.legend();fig.show()
```

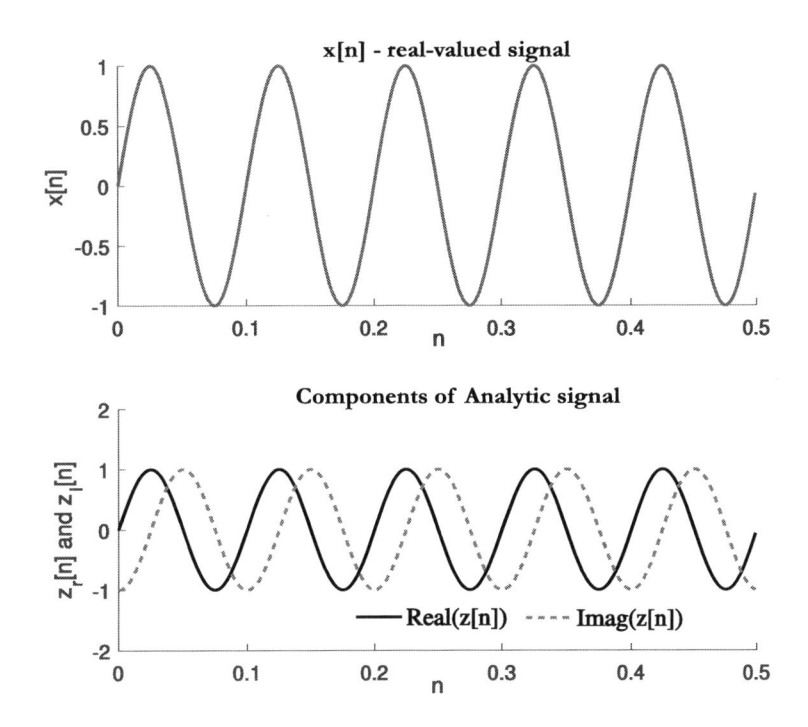

Fig. 1.22: Components of analytic signal for a real-valued sine function

1.8.1.3 Hilbert Transform using Fourier transform

SciPy's signal processing package has an inbuilt function to compute the analytic signal. The in-built function is called hilbert. We should note that the inbuilt hilbert function in SciPy returns the analytic signal $z[n]$

and not the hilbert transform of the signal $x[n]$. To get the hilbert transform, we should simply get the imaginary part of the analytic signal. Since we have written our own function to compute the analytic signal, getting the hilbert transform of a real-valued signal goes like this.

```python
import numpy as np
from essentials import analytic_signal
x_hilbert = np.imag(analytic_signal(x))
```

1.8.2 Applications of analytic signal

Hilbert transform and the concept of analytic signal has several applications. Two of them are detailed next.

1.8.2.1 Extracting instantaneous amplitude, phase, frequency

The concept of instantaneous amplitude/phase/frequency is fundamental to information communication and appears in many signal processing applications. We know that a monochromatic signal of form $x(t) = acos(\omega t + \phi)$ cannot carry any information. To carry information, the signal need to be modulated. Take for example the case of amplitude modulation, in which a positive real-valued signal $m(t)$ modulates a carrier $cos(\omega_c t)$. That is, the amplitude modulation is effected by multiplying the information bearing signal $m(t)$ with the carrier signal $cos(\omega_c t)$.

$$x(t) = m(t)cos(\omega_c t) \tag{1.80}$$

Here, ω_c is the *angular frequency* of the signal measured in radians/sec and is related to the *temporal frequency* f_c as $\omega_c = 2\pi f_c$. The term $m(t)$ is also called *instantaneous amplitude*.

Similarly, in the case of phase or frequency modulations, the concept of instantaneous phase or instantaneous frequency is required for describing the modulated signal.

$$x(t) = acos\bigl(\phi(t)\bigr) \tag{1.81}$$

Here, a is the constant amplitude factor the effects no change in the envelope of the signal and $\phi(t)$ is the instantaneous phase which varies according to the information. The *instantaneous angular frequency* is expressed as the derivative of *instantaneous phase*.

$$\omega(t) = \frac{d}{dt}\phi(t) \tag{1.82}$$

$$f(t) = \frac{1}{2\pi}\frac{d}{dt}\phi(t) \tag{1.83}$$

Generalizing these concepts, if a signal is expressed as

$$x(t) = a(t)cos\bigl(\phi(t)\bigr) \tag{1.84}$$

- The instantaneous amplitude or the *envelope* of the signal is given by $a(t)$.
- The instantaneous phase is given by $\phi(t)$.
- The instantaneous angular frequency is derived as $\omega(t) = \frac{d}{dt}\phi(t)$.
- The instantaneous temporal frequency is derived as $f(t) = \frac{1}{2\pi}\frac{d}{dt}\phi(t)$.

Let's consider the following problem statement: An amplitude modulated signal is formed by multiplying a sinusoidal information and a linear frequency chirp. The information content is expressed as $a(t) = 1+$

1.8 Analytic signal and its applications

$0.7\ sin(2\pi 3t)$ and the linear frequency chirp is made to vary from 20 Hz to 80 Hz. Given the modulated signal, extract the instantaneous amplitude (envelope), instantaneous phase and the instantaneous frequency.

The solution can be as follows. We note that the modulated signal is a real-valued signal. We also take note of the fact that amplitude, phase and frequency can be easily computed if the signal is expressed in complex form. Which transform should we use such that the we can convert a real signal to the complex plane without altering the required properties ? Answer: Apply Hilbert transform and form the analytic signal on the complex plane. Figure 1.23 illustrates this concept.

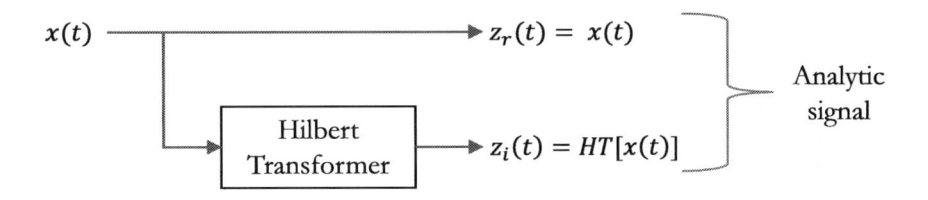

Fig. 1.23: Converting a real-valued signal to complex plane using Hilbert Transform

Express the real-valued modulated signal $x(t)$ as an analytic signal

$$z(t) = z_r(t) + jz_i(t) = x(t) + jHT\{x(t)\} \quad (1.85)$$

where, $HT\{\cdot\}$ represents the Hilbert transform operation. Now, the required parameters are very easy to obtain.

- The instantaneous amplitude is computed in the complex plane as

$$a(t) = |z(t)| = \sqrt{z_r^2(t) + z_i^2(t)} \quad (1.86)$$

- The instantaneous phase is computed in the complex plane as

$$\phi(t) = \angle z(t) = arctan\left[\frac{z_i(t)}{z_r(t)}\right] \quad (1.87)$$

- The instantaneous temporal frequency is computed in the complex plane as

$$f(t) = \frac{1}{2\pi}\frac{d}{dt}\phi(t) \quad (1.88)$$

Once we know the instantaneous phase, the carrier can be regenerated as $cos[\phi(t)]$. The regenerated carrier is often referred as *temporal fine structure (TFS)* in acoustic signal processing [16]. The following Python code demonstrates the extraction procedure and the resulting plots are shown in Figure 1.24.

Program 15: DigiCommPy\chapter_1\demo_scripts.py: Envelope and instantaneous phase extraction

```
def extract_envelope_phase():
    """
    Demonstrate extraction of instantaneous amplitude and phase from
    the analytic signal constructed from a real-valued modulated signal
    """
    import numpy as np
    from scipy.signal import chirp
    import matplotlib.pyplot as plt
    from DigiCommPy.essentials import analytic_signal
```

```python
fs = 600 # sampling frequency in Hz
t = np.arange(start=0,stop=1, step=1/fs) # time base
a_t = 1.0 + 0.7 * np.sin(2.0*np.pi*3.0*t) # information signal
c_t = chirp(t, f0=20, t1=t[-1], f1=80, phi=0, method='linear')
x = a_t * c_t # modulated signal

fig, (ax1, ax2) = plt.subplots(nrows=2, ncols=1)
ax1.plot(x) # plot the modulated signal
z = analytic_signal(x) # form the analytical signal
inst_amplitude = abs(z) # envelope extraction
inst_phase = np.unwrap(np.angle(z)) # inst phase
inst_freq = np.diff(inst_phase)/(2*np.pi)*fs # inst frequency

# Regenerate the carrier from the instantaneous phase
extracted_carrier = np.cos(inst_phase)
ax1.plot(inst_amplitude,'r') # overlay the extracted envelope
ax1.set_title('Modulated signal and extracted envelope')
ax1.set_xlabel('n');ax1.set_ylabel(r'x(t) and $|z(t)|$')
ax2.plot(extracted_carrier)
ax2.set_title('Extracted carrier or TFS')
ax2.set_xlabel('n');ax2.set_ylabel(r'$\cos[\omega(t)]$')
fig.show()
```

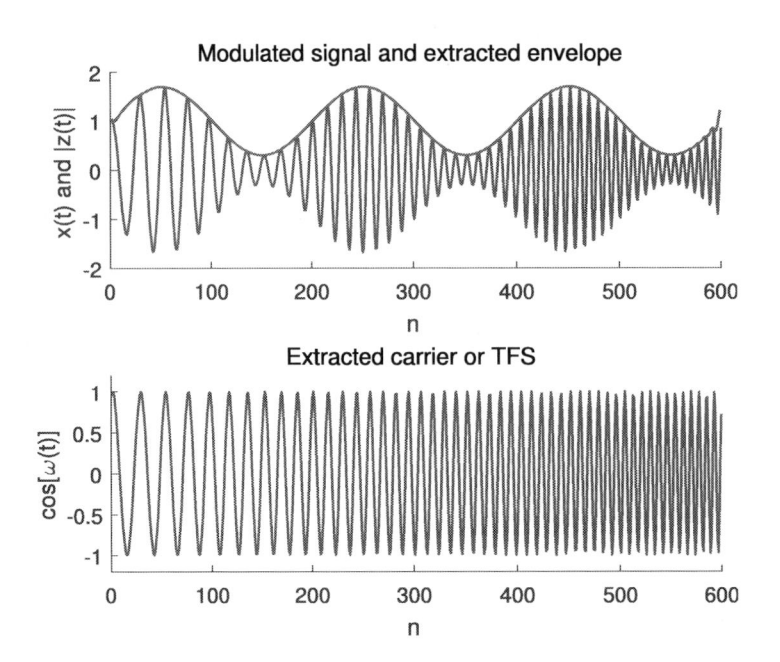

Fig. 1.24: Amplitude modulation using a chirp signal and extraction of envelope and TFS

1.8 Analytic signal and its applications

1.8.2.2 Phase demodulation using Hilbert transform

In communication systems, different types of modulations are available. They are mainly categorized as: amplitude modulation and phase modulation / frequency modulation. In amplitude modulation, the information is encoded as variations in the amplitude of a carrier signal. Demodulation of an amplitude modulated signal, involves extraction of the envelope of the modulated signal (see section 1.8.2.1).

In phase modulation, the information is encoded as variations in the phase of the carrier signal. In its generic form, a phase modulated signal is expressed as an information-bearing sinusoidal signal modulating another sinusoidal carrier signal

$$x(t) = A cos \left[2\pi f_c t + \beta + \alpha sin \left(2\pi f_m t + \theta \right) \right] \qquad (1.89)$$

where, $m(t) = \alpha sin(2\pi f_m t + \theta)$ represents the information-bearing modulating signal, with the following parameters

- α - amplitude of the modulating sinusoidal signal.
- f_m - frequency of the modulating sinusoidal signal.
- θ - phase offset of the modulating sinusoidal signal.

The carrier signal has the following parameters

- A - amplitude of the carrier.
- f_c - frequency of the carrier and $f_c >> f_m$.
- β - phase offset of the carrier.

The phase modulated signal shown in equation 1.89, can be simply expressed as

$$x(t) = A cos \left[\phi(t) \right] \qquad (1.90)$$

Here, $\phi(t)$ is the instantaneous phase that varies according to the information signal $m(t)$. A phase modulated signal of form $x(t)$ can be demodulated by forming an analytic signal (by applying Hilbert transform) and then extracting the instantaneous phase. Extraction of instantaneous phase of a signal was discussed in section 1.8.2.1.

We note that the instantaneous phase, $\phi(t) = 2\pi f_c t + \beta + \alpha sin(2\pi f_m t + \theta)$, is linear in time and proportional to $2\pi f_c t$. To obtain the information bearing modulated signal, this linear offset needs to be subtracted from the instantaneous phase. If the carrier frequency is known at the receiver, this can be done easily. If not, the carrier frequency term $2\pi f_c t$ needs to be estimated using a linear fit of the unwrapped instantaneous phase. The following Python code demonstrates all these methods. The resulting plots are shown in Figures 1.25 and 1.26.

Program 16: DigiCommPy\chapter_1\demo_scripts.py: Demodulate phase modulated signal

```python
def hilbert_phase_demod():
    """
    Demonstrate simple Phase Demodulation using Hilbert transform
    """
    import numpy as np
    import matplotlib.pyplot as plt
    from scipy.signal import hilbert
    
    fc = 210 # carrier frequency
    fm = 10 # frequency of modulating signal
    alpha = 1 # amplitude of modulating signal
    theta = np.pi/4 # phase offset of modulating signal
    beta = np.pi/5 # constant carrier phase offset
    # Set True if receiver knows carrier frequency & phase offset
```

```python
        receiverKnowsCarrier= False

        fs = 8*fc # sampling frequency
        duration = 0.5 # duration of the signal
        t = np.arange(start = 0, stop = duration, step = 1/fs) # time base

        #Phase Modulation
        m_t = alpha*np.sin(2*np.pi*fm*t + theta) # modulating signal
        x = np.cos(2*np.pi*fc*t + beta + m_t ) # modulated signal

        fig1, (ax1, ax2) = plt.subplots(nrows=2, ncols=1)
        ax1.plot(t,m_t) # plot modulating signal
        ax1.set_title('Modulating signal')
        ax1.set_xlabel('t');ax1.set_ylabel('m(t)')

        ax2.plot(t,x) # plot modulated signal
        ax2.set_title('Modulated signal')
        ax2.set_xlabel('t');ax2.set_ylabel('x(t)');fig1.show()

        #Add AWGN noise to the transmitted signal
        mu = 0; sigma = 0.1  # noise mean and sigma
        n = mu + sigma*np.random.normal(len(t)) # awgn noise
        r = x + n # noisy received signal

        # Demodulation of the noisy Phase Modulated signal
        z= hilbert(r) # form the analytical signal from the received vector
        inst_phase = np.unwrap(np.angle(z)) # instaneous phase

        if receiverKnowsCarrier: # If receiver knows the carrier freq/phase perfectly
            offsetTerm = 2*np.pi*fc*t+beta
        else: # else, estimate the subtraction term
            p = np.polyfit(x =t, y =inst_phase, deg =1)#linear fit instantaneous phase
            #re-evaluate the offset term using the fitted values
            estimated = np.polyval(p,t)
            offsetTerm = estimated

        demodulated = inst_phase - offsetTerm

        fig2, ax3 = plt.subplots()
        ax3.plot(t,demodulated) # demodulated signal
        ax3.set_title('Demodulated signal')
        ax3.set_xlabel('n')
        ax3.set_ylabel(r'$\hat{m(t)}$');fig2.show()
```

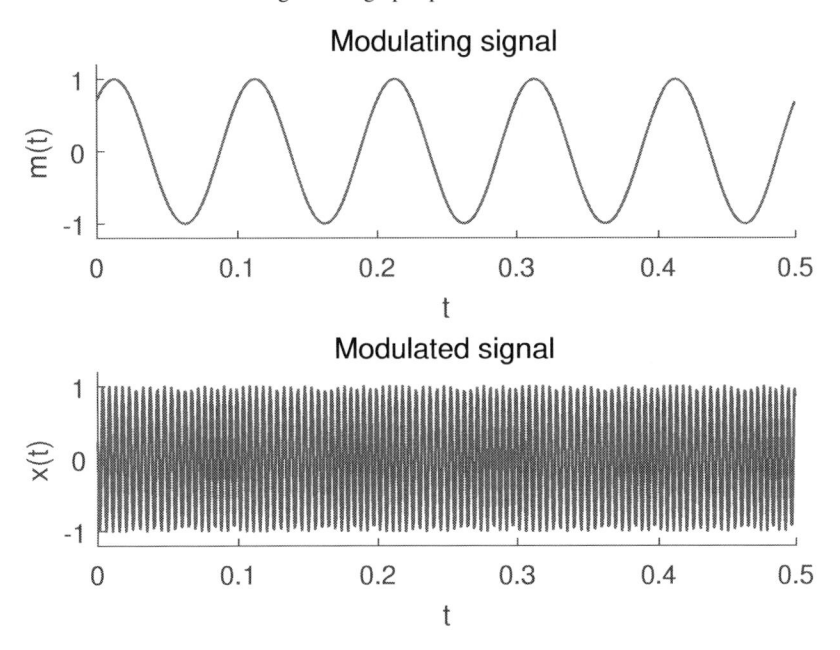

Fig. 1.25: Phase modulation - modulating signal and modulated (transmitted) signal

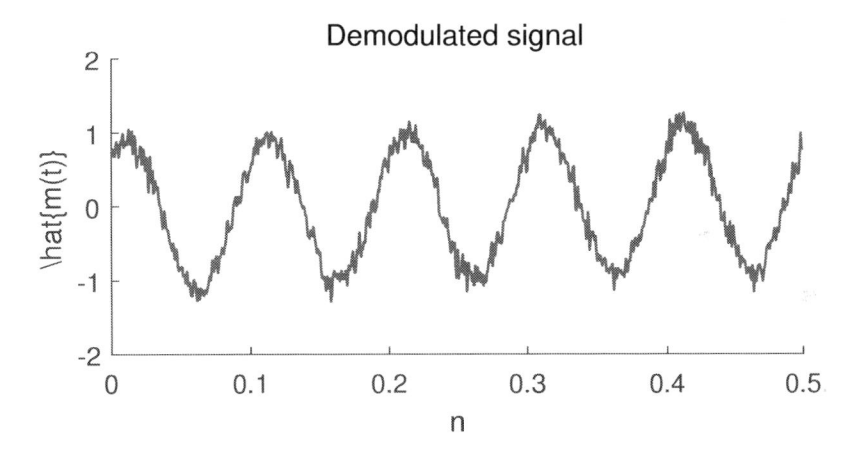

Fig. 1.26: Demodulated signal from the noisy received-signal and when the carrier is unknown at the receiver

1.9 Choosing a filter : FIR or IIR : understanding the design perspective

Choosing the best filter for implementing a signal processing block is a challenge often faced by communication systems development engineer. There exists two different types of *linear time invariant* (LTI) filters from transfer function standpoint : *finite impulse response* (FIR) and *infinite impulse response* (IIR) filters and myriad design techniques for designing them. The mere fact that there exists so many techniques for designing a filter, suggests that there is no single optimal filter design. One has to weigh-in the pros and cons of choosing a filter design by considering the factors discussed here.

1.9.1 Design specification

A filter design starts with a specification. We may have a simple specification that just calls for removing an unwanted frequency component or it can be a complete design specification that calls for various parameters like - amount of ripples allowed in passband, stop band attenuation, transition width etc. The design specification usually calls for satisfying one or more of the following:

- desired magnitude response - $|H_{spec}(\omega)|$
- desired phase response - $\angle H_{spec}(\omega)$
- tolerance specifications - that specifies how much the filter response is allowed to vary when compared with ideal response. Examples include how much ripples allowed in passband, stop band etc.

Given the specifications above, the goal of a filter design process is to choose the parameters M, N, $\{b_k\}$ and $\{a_k\}$ such that the *transfer function* of the filter

$$H(z) = \frac{\sum_{i=0}^{M} b_k z^{-1}}{1 + \sum_{i=1}^{N} a_k z^{-1}} = \frac{\prod_{i}(z-z_i)}{\prod_{j}(z-p_j)} \qquad (1.91)$$

yields the desired response: $H(\omega) \approx H_{spec}(\omega)$. In other words, the design process also involves choosing the number and location of the zeros $\{z_i\}$ and poles $\{p_j\}$ in the pole-zero plot. In Python, given the transfer function of a filter, the filtering process can be simulated using the `lfilter` function. The SciPy function `y=scipy.signal.lfilter(b,a,x)` filters the input data x using the transfer function defined by the numerator coefficients $\{b_k\}$ and denominator coefficients $\{a_k\}$.

Two types of filter can manifest from the transfer function given in equation 1.91.

- When $N = 0$, there is no feedback in the filter structure (Figure 1.27(a)), no poles in the pole-zero plot (in fact all the poles sit at the origin). The impulse response of such filter dies out (becomes zero) beyond certain point of time and it is classified as *finite impulse response (FIR)* filter. It provides *linear phase characteristic in the passband.*
- When $N > 0$, the filter structure is characterized by the presence of feedback elements (Figure 1.27(b)). Due to the presence of feedback elements, the impulse response of the filter may not become zero beyond certain point, but continues indefinitely and hence the name *infinite impulse response* (IIR) filter.
- **Caution**:In most cases, the presence of feedback elements provide infinite impulse response. It is not always true. There are some exceptional cases where the presence of feedback structure may result in finite impulse response. For example, a moving average filter will have a finite impulse response. The output of a moving average filter can be described using a recursive formula, which will result in a structure with feedback elements.

1.9.2 General considerations in design

As specified earlier, the choice of filter and the design process depends on design specification, application and the performance issues associates with them. However, the following general considerations are applied in practical design.

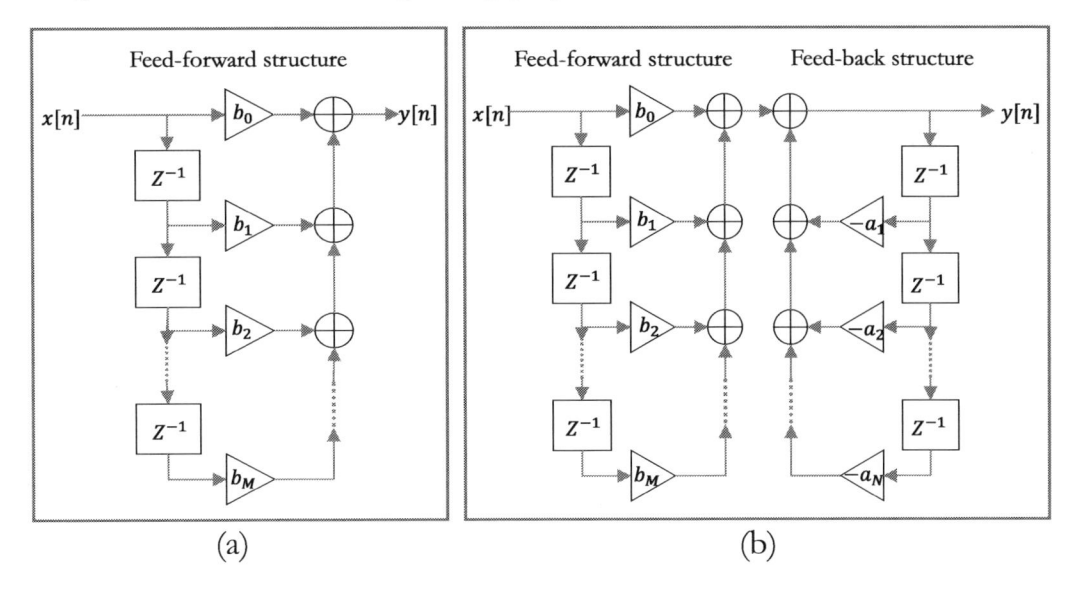

1.9 Choosing a filter : FIR or IIR : understanding the design perspective

Fig. 1.27: FIR and IIR filters can be realized as direct-form structures. (a) Structure with feed-forward elements only - typical for FIR designs. (b) Structure with feed-back paths - generally results in IIR filters. Other structures like lattice implementations are also available.

Minimizing number of computations

In order to minimize memory requirements for storing the filter co-efficients $\{a_k\},\{b_k\}$ and to minimize the number of computations, ideally we would like $N+M$ to be as small as possible. For the same specification, IIR filters result in much lower order when compared to its FIR counter part. Therefore, IIR filters are efficient when viewed from this standpoint.

Need for real-time processing

The given application may require processing of input samples in real-time or the input samples may exist in a recorded state (example: video/audio playback, image processing applications, audio compression). From this perspective, we have two types of filter systems

- Causal filter
  - Filter output depends on present and past input samples, not on the future samples. The output may also depend on the past output samples, as in IIR filters. Strictly no future samples.
  - Such filters are very much suited for real-time applications.
- Non-causal filter
  - There are many practical cases where a non-causal filter is required. Typically, such application warrants some form of post-processing, where the entire data stream is already stored in memory.
  - In such cases, a filter can be designed that can take in all type of input samples : present, past and future, for processing. These filters are classified as non-causal filters.
  - Non-causal filters have much simpler design methods.

It can be often seen in many signal processing texts, that the causal filters are practically realizable. That does not mean non-causal filters are not practically implementable. In fact both types of filters are implementable and you can see them in many systems today. The question you must ask is : whether your application requires real-time processing or processing of pre-recorded samples. If the application requires *real-time processing*, causal filters must be used. Otherwise, non-causal filters can be used.

Consequences of causal filter

If the application requires real-time processing, causal filters are the only choice for implementation. Following consequences must be considered if causality is desired. Ideal filters with finite bands of zero response (example: brick-wall filters), cannot be implemented using causal filter structure. A direct consequence of causal filter is that the response cannot be ideal. Hence, we must design the filter that provides a close approximation to the desired response . If tolerance specification is given, it has to be met. For example, in the case of designing a low pass filter with given passband frequency (ω_P) and stopband frequencies (ω_S), additional tolerance specifications like allowable passband ripple factor (δ_P), stopband ripple factor (δ_S) need to be considered for the design (Figure 1.28). Therefore, the practical filter design involves choosing ω_P, ω_S, δ_P and δ_S and then designing the filter with $N, M, \{a_k\}$ and $\{b_k\}$ that satisfies all the given requirements/responses. Often, iterative procedures may be required to satisfy all the above (example: Parks and McClellan algorithm used for designing optimal causal FIR filters [17]).

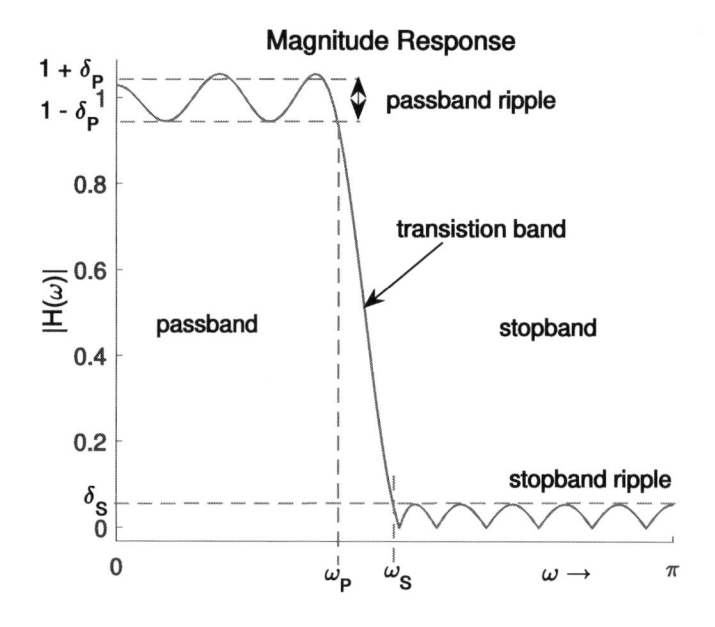

Fig. 1.28: A sample filter design specification

For a causal filter, frequency response's real part $H_R(\omega)$ and the imaginary part $H_I(\omega)$ become *Hilbert transform pair* [18]. Therefore, for a causal filter, the magnitude and phase responses become interdependent.

Stability

A causal LTI digital filter will be *bounded input bounded output* (BIBO) stable, if and only if the impulse response $h[n]$ is absolutely summable.

1.9 Choosing a filter : FIR or IIR : understanding the design perspective

$$\sum_{n=-\infty}^{n=\infty} |h[n]| < \infty \tag{1.92}$$

Impulse response of FIR filters are always bounded and hence they are inherently stable. On the other hand, an IIR filter may become unstable if not designed properly.

Consider an IIR filter implemented using a floating point processor that has enough accuracy to represent all the coefficients in the following transfer function:

$$H_1(z) = \frac{1}{1 - 1.845\, z^{-1} + 0.850586\, z^{-2}} \tag{1.93}$$

The corresponding impulse response $h_1[n]$ is plotted in Figure 1.29(a). The plot shows that the impulse response decays rapidly to zero as n increases. For this case, the sum in equation 1.92 will be finite. Hence this IIR filter is stable.

Suppose, if we were to implement the same filter in a fixed point processor and we are forced to round-off the co-efficients to 2 digits after the decimal point, the same transfer function looks like this

$$H_2(z) = \frac{1}{1 - 1.85\, z^{-1} + 0.85\, z^{-2}} \tag{1.94}$$

The corresponding impulse response $h_2[n]$ plotted in Figure 1.29(b) implies that the impulse response increases rapidly towards a constant value as n increases. For this case, the sum in equation 1.92 will approach infinity. Hence the implemented IIR filter is unstable.

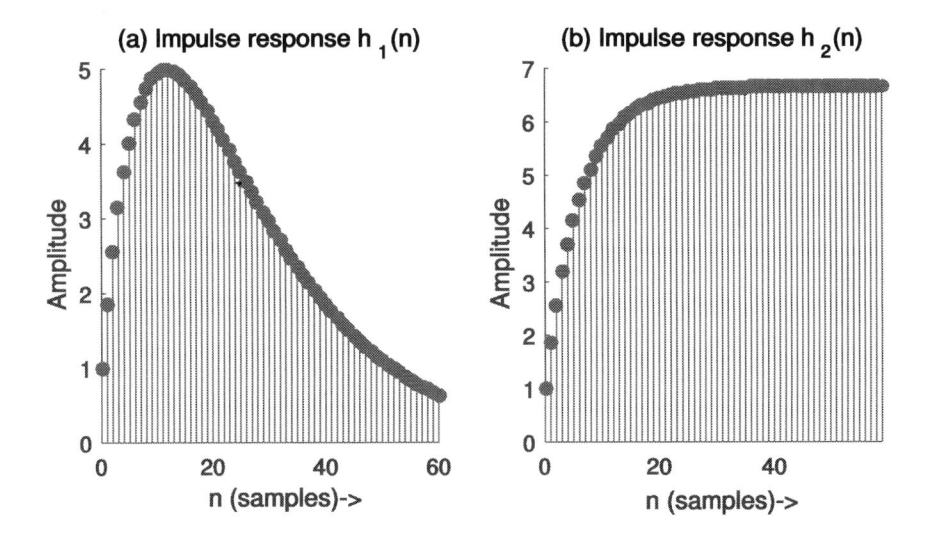

Fig. 1.29: Impact of poorly implemented IIR filter on stability. (a) Stable IIR filter, (b) The same IIR filter becomes unstable due to rounding effects.

Therefore, it is imperative that an IIR filter implementation need to be tested for stability. To analyze the stability of the filter, the infinite sum in equation 1.92 need to be computed and it is often difficult to compute this sum. Analysis of pole-zero plot is an alternative solution for this problem. To have a stable causal filter, the poles of the transfer function should rest completely, strictly, inside the unit circle on the pole-zero plot. The pole-zero plot for the above given transfer functions $H_1(z)$, $H_2(z)$ are plotted in Figure 1.30. It shows that for the transfer function $H_1(z)$, all the poles lie within the unit circle (the region of stability) and hence it is a stable IIR filter. On the other hand, for the transfer function $H_2(z)$, one pole lies exactly on the unit circle (ie., it is just out of the region of stability) and hence it is an unstable IIR filter.

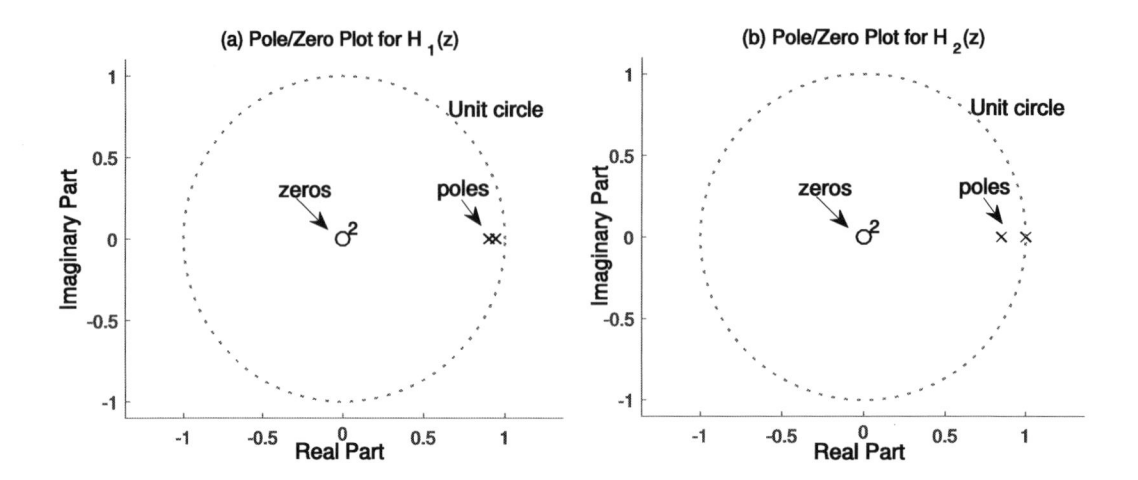

Fig. 1.30: Impact of poorly implemented IIR filter on stability. (a) Stable causal IIR filter since all poles rest inside the unit circle, (b) Due to rounding effects, one of the poles rests on the unit circle, making it unstable.

Linear phase requirement

In many signal processing applications, it is needed that a digital filter should not alter the angular relationship between the real and imaginary components of a signal, especially in the passband. In otherwords, the phase relationship between the signal components should be preserved in the filter's passband. If not, we have phase distortion.

Phase distortion is a concern in many signal processing applications. For example, in phase modulations like GMSK [19], the entire demodulation process hinges on the phase relationship between the inphase and quadrature components of the incoming signal. If we have a phase distorting filter in the demodulation chain, the entire detection process goes for a toss. Hence, we have to pay attention to the phase characteristics of such filters. To have no phase distortion when processing a signal through a filter, every spectral component of the signal inside the passband should be delayed by the same amount of time delay measured in samples. In other words, the phase response $\phi(\omega)$ in the passband should be a linear function (straight line) of frequency (except for the phase wraps at the band edges). A filter that satisfies this property is called a *linear phase filter*. FIR filters provide perfect linear phase characteristic in the passband region (Figure 1.31) and hence avoids phase distortion. All IIR filters provide non-linear phase characteristic. If a real-time application warrants for zero phase distortion, FIR filters are the immediate choice for design.

It is intuitive to see that the phase response of a generalized linear phase filter should follow the relationship $\phi(\omega) = -m\omega + c$, where m is the slope and c is the intercept when viewing the linear relationship between the frequency and the phase response in the passband (Figure 1.31). The *phase delay* and *group delay* are the important filter characteristics considered for ascertaining the phase distortion and they relate to the intercept c and the slope m of the phase response in the passband. Linear phase filters are characterized by constant group delay. Any deviation in the group delay from the constant value inside the passband, indicates presence of certain degree of non-linearity in the phase and hence causes phase distortion.

Phase delay is the time delay experienced by each spectral component of the input signal. For a filter with the frequency response $H(\omega)$, the phase delay response τ_p is defined in terms of phase response $\phi(\omega) =$

1.9 Choosing a filter : FIR or IIR : understanding the design perspective

$\angle H(\omega)$ as

$$\tau_p(\omega) = -\frac{\phi(\omega)}{\omega} \qquad (1.95)$$

Group delay is the delay experienced by a group of spectral components within a narrow frequency interval about ω [20]. The group delay response $\tau_g(\omega)$ is defined as the negative derivative of the phase response ω.

$$\tau_g(\omega) = -\frac{d}{d\omega}\phi(\omega) \qquad (1.96)$$

For the generalized linear phase filter, the group delay and phase delay are given by

$$\tau_g(\omega) = -\frac{d}{d\omega}\phi(\omega) = m \qquad (1.97)$$

$$\tau_p(\omega) = -\frac{\phi(\omega)}{\omega} = m - \frac{c}{\omega} \qquad (1.98)$$

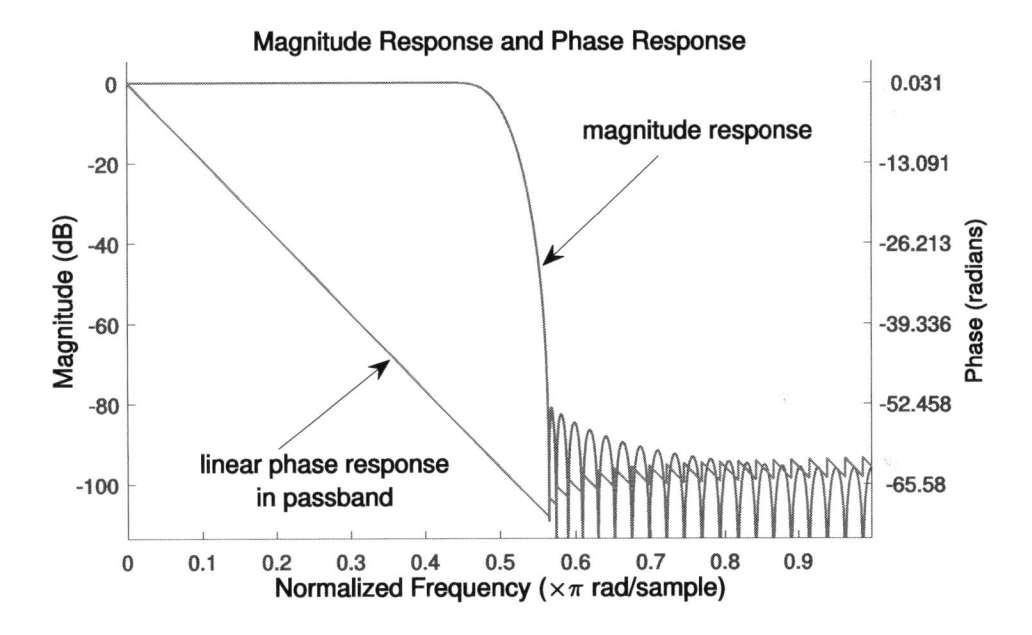

Fig. 1.31: An FIR filter showing linear phase characteristic in the passband

Summary of design choices

- IIR filters are efficient, they can provide similar magnitude response for fewer coefficients or lower side-lobes for same number of coefficients
- For linear phase requirement, FIR filters are the immediate choice for the design
- FIR filters are inherently stable. IIR filters are susceptible to finite length words effects of fixed point arithmetic and hence the design has to be rigorously tested for stability.

- IIR filters provide less average delay compared to its equivalent FIR counterpart. If the filter has to be used in a feedback path in a system, the amount of filter delay is very critical as it affects the stability of the overall system.
- Given a specification, an IIR design can be easily deduced based on closed-form expressions. However, satisfying the design requirements using an FIR design, generally requires iterative procedures.

References

1. James W. Cooley and John W. Tukey, *An Algorithm for the Machine Calculation of Complex Fourier Series*, Mathematics of Computation Vol. 19, No. 90, pp. 297-301, April 1965.
2. Good I. J, *The interaction algorithm and practical Fourier analysis*, Journal of the Royal Statistical Society, Series B, Vol. 20, No. 2, pp. 361-372, 1958.
3. Thomas L. H, *Using a computer to solve problems in physics*, Applications of Digital Computers, Boston, Ginn, 1963.
4. *Numpy.spacing*, Numpy v1.16 manual, https://docs.scipy.org/doc/numpy/reference/generated/numpy.spacing.html
5. Hanspeter Schmid, *How to use the FFT and Matlab's pwelch function for signal and noise simulations and measurements*, Institute of Microelectronics, University of Applied Sciences Northwestern Switzerland, August 2012.
6. Sanjay Lall, Norm and Vector spaces, Information Systems Laboratory, Stanford. University.http://floatium.stanford.edu/engr207c/lectures/norms_2008_10_07_01.pdf
7. Reddi.S.S, *Eigen Vector properties of Toeplitz matrices and their application to spectral analysis of time series*, Signal Processing, Vol 7, North-Holland, 1984, pp. 46-56.
8. Robert M. Gray, *Toeplitz and circulant matrices – an overview*, Department of Electrical Engineering, Stanford University, Stanford 94305,USA. https://ee.stanford.edu/~gray/toeplitz.pdf
9. SciPy documentation help on Toeplitz function. https://docs.scipy.org/doc/scipy-0.14.0/reference/generated/scipy.linalg.toeplitz.html
10. Richard E. Blahut, *Fast Algorithms for Signal Processing*, Cambridge University Press, 1 edition, August 2010.
11. D. Gabor, *Theory of communications*, Journal of the Inst. Electr. Eng., vol. 93, pt. 111, pp. 42-57, 1946. See definition of complex signal on p. 432.
12. J. A. Ville, *Theorie et application de la notion du signal analytique*, Cables el Transmission, vol. 2, pp. 61-74, 1948.
13. S. M. Kay, *Maximum entropy spectral estimation using the analytical signal*, IEEE transactions on Acoustics, Speech, and Signal Processing, vol. 26, pp. 467-469, October 1978.
14. Poularikas A. D. *Handbook of Formulas and Tables for Signal Processing*, ISBN - 9781420049701, CRC Press, 1999.
15. S. L. Marple, *Computing the discrete-time analytic signal via FFT*, Conference Record of the Thirty-First Asilomar Conference on Signals, Systems and Computers, Pacific Grove, CA, USA, 1997, pp. 1322-1325 vol.2.
16. Moon, Il Joon, and Sung Hwa Hong. *What Is Temporal Fine Structure and Why Is It Important ?*, Korean Journal of Audiology 18.1 (2014): 1–7. PMC. Web. 24 Apr. 2017.
17. J. H. McClellan, T. W. Parks, and L. R. Rabiner, *A Computer Program for Designing Optimum FIR Linear Phase Digital Filters*, IEEE Trans, on Audio and Electroacoustics, Vol. AU-21, No. 6, pp. 506-526, December 1973.
18. Frank R. Kschischang, *The Hilbert Transform*, Department of Electrical and Computer Engineering, University of Toronto, October 22, 2006.
19. Thierry Turletti, *GMSK in a nutshell*, Telemedia Networks and Systems Group Laboratory for Computer Science, Massachussets Institute of Technology April, 1996.
20. Julius O. Smith III, *Introduction to digital filters - with audio applications*, Center for Computer Research in Music and Acoustics (CCRMA), Department of Music, Stanford University.

Chapter 2
Digital Modulators and Demodulators - Passband Simulation Models

Abstract This chapter focuses on the passband simulation models for various modulation techniques including BPSK, differentially encoded BPSK, differential BPSK, QPSK, offset QPSK, $\pi/4$-DQPSK, CPM and MSK, GMSK, FSK. Power spectral density (PSD) and performance analysis for these techniques are also provided.

2.1 Introduction

For a given modulation technique, there are two ways to implement the simulation model: *passband model* and *equivalent baseband model*. The passband model is also called *waveform level simulation model*. The waveform level simulation techniques, described in this chapter, are used to represent the physical interactions of the transmitted signal with the channel. In the waveform level simulations, the transmitted signal, the noise and received signal are all represented by samples of waveforms.

Typically, a waveform level simulation uses many samples per symbol. For the computation of error rate performance of various digital modulation techniques, the value of the symbol at the symbol-sampling time instant is all the more important than the look of the entire waveform. In such a case, the detailed waveform level simulation is not required, instead *equivalent baseband discrete-time model*, described in chapter 3 can be used. Discrete-time equivalent channel model requires only one sample per symbol, hence it consumes less memory and yields results in a very short span of time.

In any communication system, the transmitter operates by modulating the information bearing baseband waveform on to a sinusoidal RF carrier resulting in a *passband* signal. The carrier frequency, chosen for transmission, varies for different applications. For example, FM radio uses $88 - 108$ *MHz* carrier frequency range, whereas for indoor wireless networks the center frequency of transmission is 1.8 *GHz*. Hence, the carrier frequency is not the component that contains the information, rather it is the baseband signal that contains the information that is being conveyed.

Actual RF transmission begins by converting the baseband signals to passband signals by the process of up-conversion. Similarly, the passband signals are down-converted to baseband at the receiver, before actual demodulation could begin. Based on this context, two basic types of behavioral models exist for simulation of communication systems - passband models and its baseband equivalent. In the passband model, every cycle of the RF carrier is simulated in detail and the power spectrum will be concentrated near the carrier frequency f_c. Hence, passband models consume more memory, as every point in the RF carrier needs to be stored in computer memory for simulation.

On the other hand, the signals in baseband models are centered near zero frequency. In baseband equivalent models, the RF carrier is suppressed and therefore the number of samples required for simulation is greatly reduced. Furthermore, if the behavior of the system is well understood, the baseband model can be further simplified and the system can be implemented entirely based on the samples at symbol-sampling time instants. Conversion of passband model to baseband equivalent model is discussed in chapter 3 section 3.2.

2.2 Binary Phase Shift Keying (BPSK)

Binary Phase Shift Keying (BPSK) is a two phase modulation scheme, where the 0's and 1's in a binary message are represented by two different phase states in the carrier signal: $\theta = 0°$ for binary 1 and $\theta = 180°$ for binary 0.

In digital modulation techniques, a set of basis functions are chosen for a particular modulation scheme. Generally, the basis functions are orthogonal to each other. Basis functions can be derived using *Gram Schmidt orthogonalization* procedure [1]. Once the basis functions are chosen, any vector in the signal space can be represented as a linear combination of them. In BPSK, only one sinusoid is taken as the basis function. Modulation is achieved by varying the phase of the sinusoid depending on the message bits. Therefore, within a bit duration T_b, the two different phase states of the carrier signal are represented as

$$\begin{aligned} s_1(t) &= A_c \, cos(2\pi f_c t), & 0 \leq t \leq T_b \quad \text{for binary 1} \\ s_0(t) &= A_c \, cos(2\pi f_c t + \pi), & 0 \leq t \leq T_b \quad \text{for binary 0} \end{aligned} \quad (2.1)$$

where, A_c is the amplitude of the sinusoidal signal, f_c is the carrier frequency (Hz), t being the instantaneous time in seconds, T_b is the bit period in seconds. The signal $s_0(t)$ stands for the carrier signal when information bit $a_k = 0$ was transmitted and the signal $s_1(t)$ denotes the carrier signal when information bit $a_k = 1$ was transmitted.

The constellation diagram for BPSK (Figure 2.3) will show two constellation points, lying entirely on the x axis (inphase). It has no projection on the y axis (quadrature). This means that the BPSK modulated signal will have an in-phase component but no quadrature component. This is because it has only one basis function. It can be noted that the carrier phases are 180° apart and it has constant envelope. The carrier's phase contains all the information that is being transmitted.

2.2.1 BPSK transmitter

A BPSK transmitter, shown in Figure 2.1, is implemented by coding the message bits using NRZ coding (1 represented by positive voltage and 0 represented by negative voltage) and multiplying the output by a reference oscillator running at carrier frequency f_c.

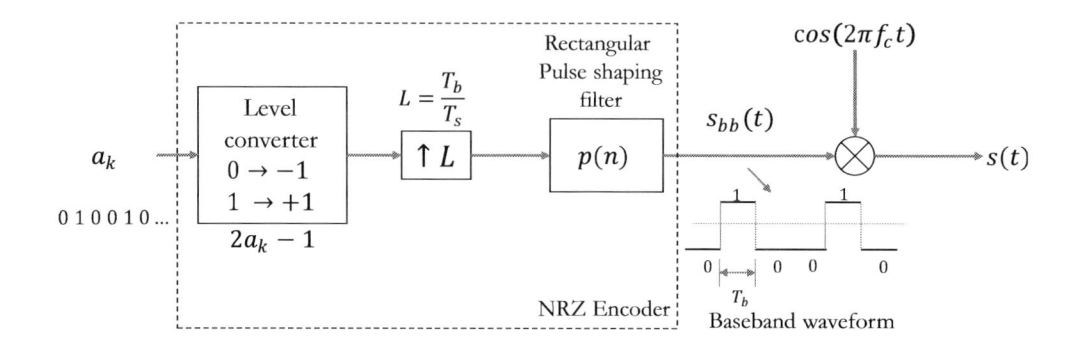

Fig. 2.1: BPSK transmitter

2.2 Binary Phase Shift Keying (BPSK)　　　　　　　　　　　　　　　　　　　　　　　　　57

The following function (bpsk_mod) implements a *baseband* BPSK transmitter according to Figure 2.1. The output of the function is in baseband and it can optionally be multiplied with the carrier frequency outside the function. In order to get nice continuous curves, the oversampling factor (*L*) in the simulation should be appropriately chosen. If a carrier signal is used, it is convenient to choose the oversampling factor as the ratio of sampling frequency (f_s) and the carrier frequency (f_c). The chosen sampling frequency must satisfy the Nyquist sampling theorem with respect to carrier frequency. For baseband waveform simulation, the oversampling factor can simply be chosen as the ratio of bit period (T_b) to the chosen sampling period (T_s), where the sampling period is sufficiently smaller than the bit period.

Program 17: DigiCommPy\passband_modulations.py: Baseband BPSK modulator

```
def bpsk_mod(ak,L):
    """
    Function to modulate an incoming binary stream using BPSK (baseband)
    Parameters:
        ak : input binary data stream (0's and 1's) to modulate
        L : oversampling factor (Tb/Ts)
    Returns:
        (s_bb,t) : tuple of following variables
                s_bb: BPSK modulated signal(baseband) - s_bb(t)
                t :  generated time base for the modulated signal
    """
    from scipy.signal import upfirdn
    s_bb = upfirdn(h=[1]*L, x=2*ak-1, up = L) # NRZ encoder
    t=np.arange(start = 0,stop = len(ak)*L) #discrete time base
    return (s_bb,t)
```

2.2.2 BPSK receiver

A correlation type coherent detector, shown in Figure 2.2, is used for receiver implementation. In coherent detection technique, the knowledge of the carrier frequency and phase must be known to the receiver. This can be achieved by using a *Costas loop* or a *Phase Lock Loop (PLL)* at the receiver. For simulation purposes, we simply assume that the carrier phase recovery was done and therefore we directly use the generated reference frequency at the receiver - $cos(2\pi f_c t)$.

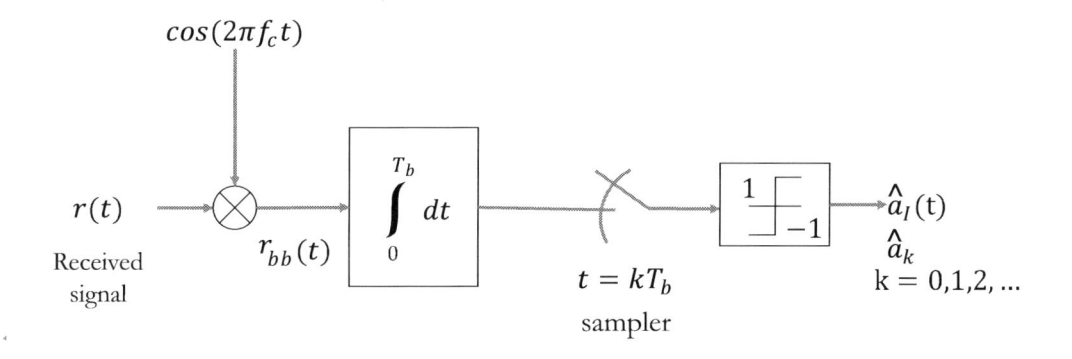

Fig. 2.2: Coherent detection of BPSK (correlation type)

In the coherent receiver, the received signal is multiplied by a reference frequency signal from the carrier recovery blocks like PLL or Costas loop. Here, it is assumed that the PLL/Costas loop is present and the output is completely synchronized. The multiplied output is integrated over one bit period using an integrator. A threshold detector makes a decision on each integrated bit based on a threshold. Since, NRZ signaling format was used in the transmitter, the threshold for the detector would be set to 0. The function bpsk_demod, implements a *baseband* BPSK receiver according to Figure 2.2. To use this function in waveform simulation, first, the received waveform has to be downconverted to baseband, and then the function may be called.

Program 18: DigiCommPy\passband_modulations.py: Baseband BPSK detection

```python
def bpsk_demod(r_bb,L):
    """
    Function to demodulate a BPSK (baseband) signal
    Parameters:
        r_bb : received signal at the receiver front end (baseband)
        L : oversampling factor (Tsym/Ts)
    Returns:
        ak_hat : detected/estimated binary stream
    """
    x = np.real(r_bb) # I arm
    x = np.convolve(x,np.ones(L)) # integrate for Tb duration (L samples)
    x = x[L-1:-1:L] # I arm - sample at every L
    ak_hat = (x > 0).transpose() # threshold detector
    return ak_hat
```

2.2.3 End-to-end simulation

The complete waveform simulation for the end-to-end transmission of information using BPSK modulation is given next. The simulation involves: generating random message bits, modulating them using BPSK modulation, addition of AWGN noise according to the chosen signal-to-noise ratio and demodulating the noisy signal using a coherent receiver. The topic of adding AWGN noise according to the chosen signal-to-noise ratio is discussed in section 4.1 in chapter 4.

The resulting waveform plots are shown in the Figure 2.3. The performance simulation for the BPSK transmitter/receiver combination is also coded in the program shown next (see chapter 4 for more details on theoretical error rates). The resulting performance curves will be same as the ones obtained using the complex baseband equivalent simulation technique in Figure 4.4 of chapter 4.

Program 19: DigiCommPy\chapter_2\bpsk.py: Performance of BPSK using waveform simulation

```python
#Execute in Python3: exec(open("chapter_2/bpsk.py").read())
import numpy as np #for numerical computing
import matplotlib.pyplot as plt #for plotting functions
from DigiCommPy.passband_modulations import bpsk_mod, bpsk_demod
from DigiCommPy.channels import awgn
from scipy.special import erfc

N=100000 # Number of symbols to transmit
EbN0dB = np.arange(start=-4,stop = 11,step = 2) # Eb/N0 range in dB for simulation
L=16 # oversampling factor,L=Tb/Ts(Tb=bit period,Ts=sampling period)
```

2.3 Coherent detection of Differentially Encoded BPSK (DEBPSK)

```python
# if a carrier is used, use L = Fs/Fc, where Fs >> 2xFc
Fc=800 # carrier frequency
Fs=L*Fc # sampling frequency
BER = np.zeros(len(EbN0dB)) # for BER values for each Eb/N0
ak = np.random.randint(2, size=N) # uniform random symbols from 0's and 1's
(s_bb,t)= bpsk_mod(ak,L) # BPSK modulation(waveform) - baseband
s = s_bb*np.cos(2*np.pi*Fc*t/Fs) # with carrier
# Waveforms at the transmitter
fig1, axs = plt.subplots(2, 2)
axs[0, 0].plot(t,s_bb) # baseband wfm zoomed to first 10 bits
axs[0, 0].set_xlabel('t(s)');axs[0, 1].set_ylabel(r'$s_{bb}(t)$-baseband')
axs[0, 1].plot(t,s) # transmitted wfm zoomed to first 10 bits
axs[0, 1].set_xlabel('t(s)');axs[0, 1].set_ylabel('s(t)-with carrier')
axs[0, 0].set_xlim(0,10*L);axs[0, 1].set_xlim(0,10*L)
#signal constellation at transmitter
axs[1, 0].plot(np.real(s_bb),np.imag(s_bb),'o')
axs[1, 0].set_xlim(-1.5,1.5);axs[1, 0].set_ylim(-1.5,1.5)

for i,EbN0 in enumerate(EbN0dB):
    # Compute and add AWGN noise
    r = awgn(s,EbN0,L) # refer Chapter section 4.1
    
    r_bb = r*np.cos(2*np.pi*Fc*t/Fs) # recovered baseband signal
    ak_hat = bpsk_demod(r_bb,L) # baseband correlation demodulator
    BER[i] = np.sum(ak !=ak_hat)/N # Bit Error Rate Computation
    
    # Received signal waveform zoomed to first 10 bits
    axs[1, 1].plot(t,r) # received signal (with noise)
    axs[1, 1].set_xlabel('t(s)');axs[1, 1].set_ylabel('r(t)')
    axs[1, 1].set_xlim(0,10*L)
#------Theoretical Bit/Symbol Error Rates-------------
theoreticalBER = 0.5*erfc(np.sqrt(10**(EbN0dB/10))) # Theoretical bit error rate
#-------------Plots--------------------------
fig2, ax1 = plt.subplots(nrows=1,ncols = 1)
ax1.semilogy(EbN0dB,BER,'k*',label='Simulated') # simulated BER
ax1.semilogy(EbN0dB,theoreticalBER,'r-',label='Theoretical')
ax1.set_xlabel(r'$E_b/N_0$ (dB)')
ax1.set_ylabel(r'Probability of Bit Error - $P_b$')
ax1.set_title(['Probability of Bit Error for BPSK modulation'])
ax1.legend();fig1.show();fig2.show()
```

2.3 Coherent detection of Differentially Encoded BPSK (DEBPSK)

In coherent detection, the receiver derives its demodulation frequency and phase references using a carrier synchronization loop. Such synchronization circuits may introduce phase ambiguity $\phi = \hat{\theta} - \theta$ in the detected phase, which could lead to erroneous decisions in the demodulated bits. For example, Costas loop exhibits phase ambiguity of integral multiples of π radians at the lock-in points. As a consequence, the carrier recovery

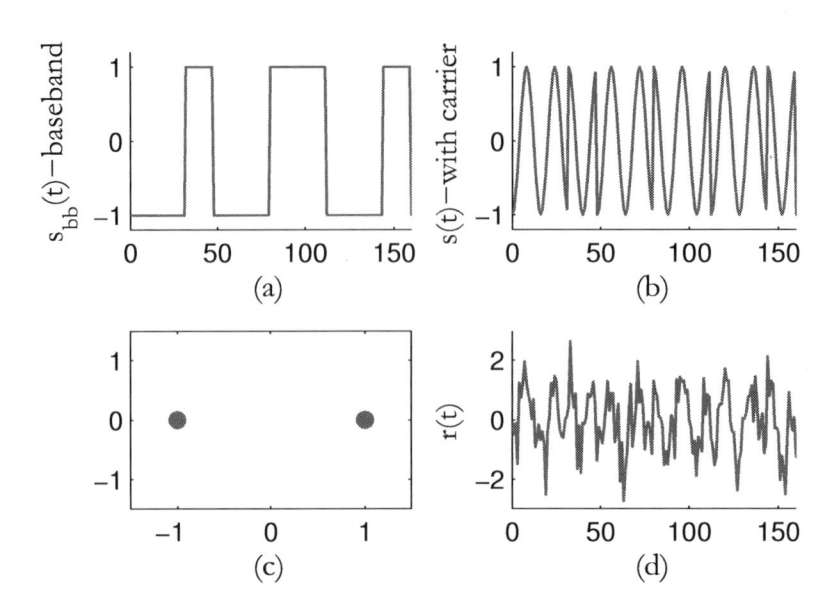

Fig. 2.3: (a) Baseband BPSK signal,(b) transmitted BPSK signal - with carrier, (c) constellation at transmitter and (d) received signal with AWGN noise

may lock in π radians out-of-phase thereby leading to a situation where all the detected bits are completely inverted when compared to the bits during perfect carrier synchronization. Phase ambiguity can be efficiently combated by applying differential encoding at the BPSK modulator input (Figure 2.4) and by performing differential decoding at the output of the coherent demodulator at the receiver side (Figure 2.5).

In ordinary BPSK transmission, the information is encoded as absolute phases: $\theta = 0°$ for binary 1 and $\theta = 180°$ for binary 0. With differential encoding, the information is encoded as the phase difference between two successive samples. Assuming $a[k]$ is the message bit intended for transmission, the differential encoded output is given as

$$b[k] = b[k-1] \oplus a[k] \quad (modulo-2) \tag{2.2}$$

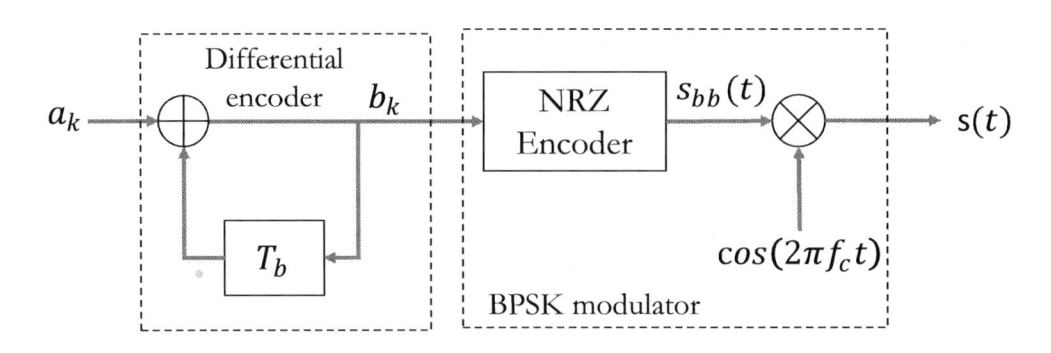

Fig. 2.4: Differential encoded BPSK transmission

The differentially encoded bits are then BPSK modulated and transmitted. On the receiver side, the BPSK sequence is coherently detected and then decoded using a differential decoder. The differential decoding is mathematically represented as

2.3 Coherent detection of Differentially Encoded BPSK (DEBPSK)

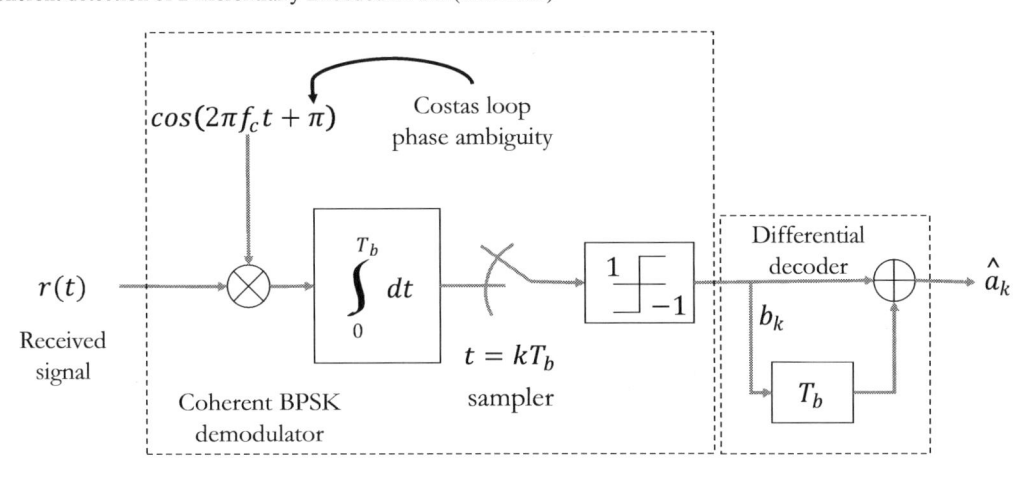

Fig. 2.5: Coherent detection of differentially encoded BPSK signal

$$a[k] = b[k] \oplus b[k-1] \quad (modulo-2) \tag{2.3}$$

This method can deal with the 180° phase ambiguity introduced by synchronization circuits. However, it suffers from performance penalty due to the fact that the differential decoding produces wrong bits when: a) the preceding bit is in error and the present bit is not in error, or b) when the preceding bit is not in error and the present bit is in error. The average bit error rate of coherently detected differentially encoded BPSK over AWGN channel is given by

$$P_b = erfc\left(\sqrt{\frac{E_b}{N_0}}\right)\left[1 - \frac{1}{2}erfc\left(\sqrt{\frac{E_b}{N_0}}\right)\right] \tag{2.4}$$

Following is the Python implementation of the waveform simulation model for the method discussed here. Both the differential encoding and differential decoding blocks, illustrated in Figures 2.4 and 2.5), are linear time-invariant filters. The differential encoder is realized using IIR type digital filter and the differential decoder is realized as FIR filter.

Program 20: DigiCommPy\chapter_2\debpsk_coherent.py: Coherent detection of DEBPSK

```python
#Execute in Python3: exec(open("chanter_2/debpsk_coherent.py").read())
import numpy as np #for numerical computing
import matplotlib.pyplot as plt #for plotting functions
from DigiCommPy.passband_modulations import bpsk_mod, bpsk_demod
from DigiCommPy.channels import awgn
from scipy.signal import lfilter
from scipy.special import erfc

N=1000000 # Number of symbols to transmit
EbN0dB = np.arange(start=-4,stop = 11,step = 2) # Eb/N0 range in dB for simulation
L=16 # oversampling factor,L=Tb/Ts(Tb=bit period,Ts=sampling period)
# if a carrier is used, use L = Fs/Fc, where Fs >> 2xFc
Fc=800 # carrier frequency
Fs=L*Fc # sampling frequency
```

```python
SER = np.zeros(len(EbN0dB)) # for SER values for each Eb/N0

ak = np.random.randint(2, size=N) # uniform random symbols from 0's and 1's
bk = lfilter([1.0],[1.0,-1.0],ak) #IIR filter for differential encoding
bk = bk%2 #XOR operation is equivalent to modulo-2

[s_bb,t]= bpsk_mod(bk,L) # BPSK modulation(waveform) - baseband
s = s_bb*np.cos(2*np.pi*Fc*t/Fs) # DEBPSK with carrier

for i,EbN0 in enumerate(EbN0dB):
    # Compute and add AWGN noise
    r = awgn(s,EbN0,L) # refer Chapter section 4.1

    phaseAmbiguity=np.pi # 180* phase ambiguity of Costas loop
    r_bb=r*np.cos(2*np.pi*Fc*t/Fs+phaseAmbiguity) # recovered signal
    b_hat=bpsk_demod(r_bb,L) # baseband correlation type demodulator
    a_hat=lfilter([1.0,1.0],[1.0],b_hat) # FIR for differential decoding
    a_hat= a_hat % 2 # binary messages, therefore modulo-2
    SER[i] = np.sum(ak !=a_hat)/N #Symbol Error Rate Computation
#------Theoretical Bit/Symbol Error Rates-------------
EbN0lins = 10**(EbN0dB/10) # converting dB values to linear scale
theorySER_DPSK = erfc(np.sqrt(EbN0lins))*(1-0.5*erfc(np.sqrt(EbN0lins)))
theorySER_BPSK = 0.5*erfc(np.sqrt(EbN0lins))
#-------------Plots------------------------
fig, ax = plt.subplots(nrows=1,ncols = 1)
ax.semilogy(EbN0dB,SER,'k*',label='Coherent DEBPSK(sim)')
ax.semilogy(EbN0dB,theorySER_DPSK,'r-',label='Coherent DEBPSK(theory)')
ax.semilogy(EbN0dB,theorySER_BPSK,'b-',label='Conventional BPSK')
ax.set_title('Probability of Bit Error for BPSK over AWGN');
ax.set_xlabel(r'$E_b/N_0$ (dB)');ax.set_ylabel(r'Probability of Bit Error - $P_b$');
ax.legend();fig.show()
```

Figure 2.6 shows the simulated BER points together with the theoretical BER curves for differentially encoded BPSK and the conventional coherently detected BPSK system over AWGN channel.

2.4 Differential BPSK (D-BPSK)

In section 2.3, differential encoding and coherent detection of binary data was discussed. It was designated as differentially encoded BPSK (DEBPSK). DEBPSK signal can be coherently detected or differentially demodulated (non-coherent detection). In a DEBPSK signal, the information is encoded as the phase difference between two successive samples. The coherent detection method for DEBPSK has not made use of this advantage. If this information can be exploited, the DEBPSK signal can be non-coherently detected.

The differential demodulation of DEBPSK signal is denoted as differential-BPSK or DBPSK, sometimes simply called as DPSK.

2.4 Differential BPSK (D-BPSK)

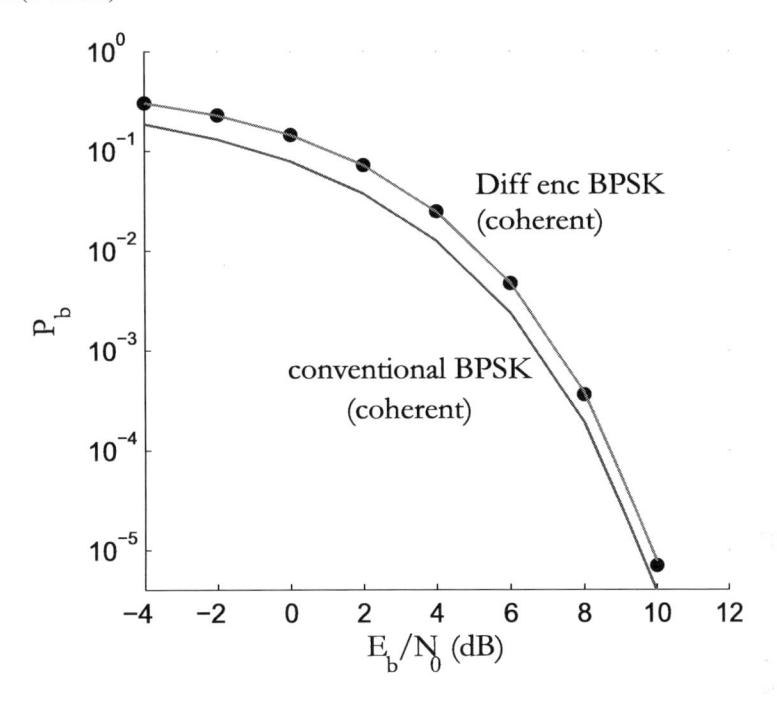

Fig. 2.6: Performance of differentially encoded BPSK and conventional BPSK over AWGN channel

2.4.1 Sub-optimum receiver for DBPSK

Coherent detection requires a reference signal with accurate frequency and phase information. However, in non-coherent communications, a coherent reference signal is not available at the receiver. Figure 2.7 illustrates a sub-optimum receiver, which differentially demodulates the DBPSK signal, where the previous symbol serves as a reference to demodulate the current symbol.

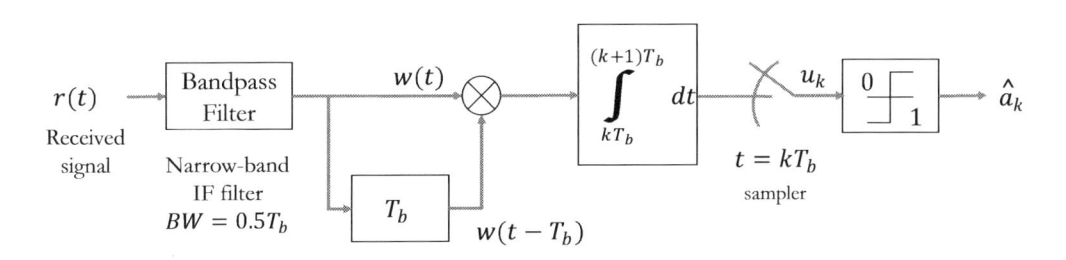

Fig. 2.7: DBPSK sub-optimum Receiver

In the sub-optimum receiver, the received signal is first passed through a narrow-band intermediate (IF) filter with bandwidth set to $BW = 0.5T_b$, where T_b is the bit duration. The IF filter not only reduces the noise in the received signal but also preserves the phase relationships contained in the signal. The filtered output is used to generate a differential signal that drives the integrator. The output of the integrator is given by

$$z = \int_{kT_b}^{(k+1)T_b} r(t)r(t-T_b)dt \tag{2.5}$$

In the absence of noise, the output of the integrator becomes

$$z = \int_{kT_b}^{(k+1)T_b} s_k(t)s_{k-1}(t)dt = \begin{cases} E_b & , if\ s_k(t) = s_{k-1}(t) \\ -E_b & , if\ s_k(t) = -s_{k-1}(t) \end{cases} \quad (2.6)$$

where $s_k(t)$ and $s_{k-1}(t)$ are the current and previous transmitted symbols. Thus, the receiver makes decisions based on the difference between two adjacent bits. As a result, a differentially encoded BPSK (DEBPSK) signal can be differentially demodulated using the method above and this receiver structure is termed as *sub-optimum DBPSK receiver*. The bit error probability performance of the sub-optimum receiver is given by [2]

$$P_b = \frac{1}{2}exp\left(-0.76\frac{E_b}{N_0}\right) \quad (2.7)$$

2.4.2 Optimum non-coherent receiver for DBPSK

A simpler form of optimum non-coherent receiver for DBPSK is given in Figure 2.8 and its derivation can be found in [3]. In differentially encoded BPSK (DEBPSK), each message bit is represented by two adjacent modulated symbols. If the transmitted bit is 0, the two modulated symbols are same and if the transmitted bit is 1, the two modulated symbols will be different. Since, the phase information is encoded in two adjacent modulated symbols, following signals can be defined for $2T_b$ duration.

$$\xi_1 = \begin{cases} A\cos(2\pi f_c t), & 0 \le t \le T_b \\ A\cos(2\pi f_c t), & T_b \le t \le 2T_b \end{cases} \text{, for binary 0} \quad (2.8)$$

$$\xi_2 = \begin{cases} A\cos(2\pi f_c t), & 0 \le t \le T_b \\ -A\cos(2\pi f_c t), & T_b \le t \le 2T_b \end{cases} \text{, for binary 1} \quad (2.9)$$

Then the decision rule is given by

$$a_k \triangleq w_k w_{k-1} + z_k z_{k-1} \underset{1}{\overset{0}{\gtrless}} 0 \quad (2.10)$$

where, the signals w_k, w_{k-1}, z_k and z_{k-1} are defined for the bit $2T_b$ duration as

$$w_0 = \int_0^T r(t)A\cos(2\pi f_c t)dt$$
$$w_1 = \int_T^{2T} r(t)A\cos(2\pi f_c t)dt$$
$$z_0 = \int_0^T r(t)A\sin(2\pi f_c t)dt$$
$$z_1 = \int_T^{2T} r(t)A\sin(2\pi f_c t)dt \quad (2.11)$$

This method does not require phase synchronization between the transmitter and receiver. However, it does require an accurate frequency reference for the carrier. Any change in the carrier frequency must be tracked and synchronized. Therefore, the sub-optimum receiver is a more practical, less complex implementation but is slightly inferior to the optimum receiver. The error rate performance of the optimum non-coherent DBPSK receiver is given by

$$P_b = \frac{1}{2}exp\left(-\frac{E_b}{N_0}\right) \quad (2.12)$$

2.4 Differential BPSK (D-BPSK)

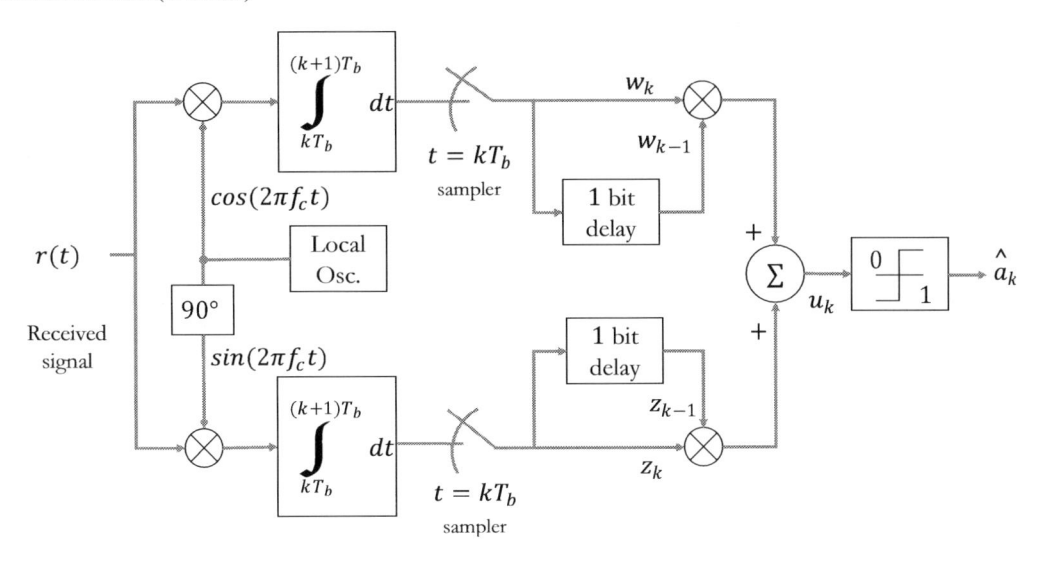

Fig. 2.8: DBPSK optimum non-coherent Receiver

The Python implementation of DEBPSK modulation and its demodulation using both the sub-optimum receiver (Figure 2.7) and the optimum receiver (Figure 2.8) techniques are given next. Figure 2.9 shows the simulated points for the sub-optimum and the optimum receiver for DBPSK, together with the theoretical curves for conventional BPSK and coherently demodulated DEBPSK schemes.

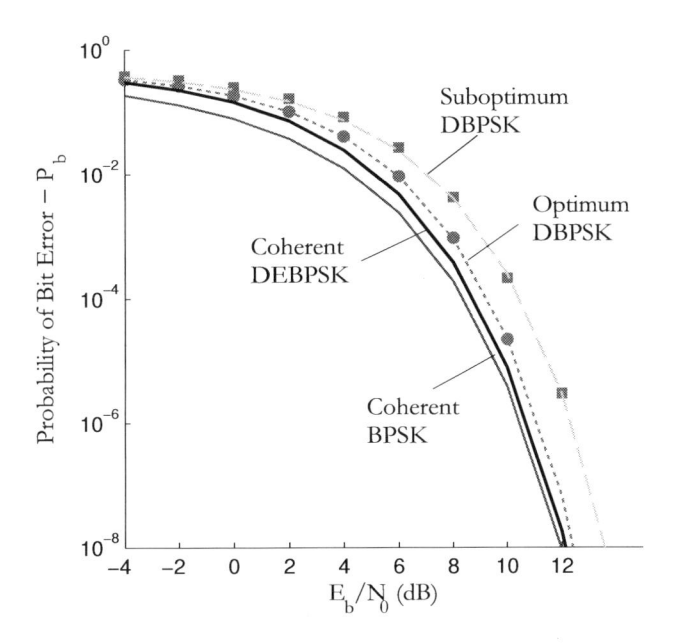

Fig. 2.9: Performance of differential BPSK schemes and coherently detected conventional BPSK

Program 21: DigiCommPy\chapter_2\dbpsk_noncoherent.py: DBPSK non-coherent detection

```python
#Execute in Python3: exec(open("chapter_2/dbpsk_noncoherent.py").read())
import numpy as np #for numerical computing
import matplotlib.pyplot as plt #for plotting functions
from DigiCommPy.passband_modulations import bpsk_mod
from DigiCommPy.channels import awgn
from scipy.signal import lfilter
from scipy.special import erfc

N=100000 # Number of symbols to transmit
EbN0dB = np.arange(start=-4,stop = 11,step = 2) # Eb/N0 range in dB for simulation
L=8 # oversampling factor,L=Tb/Ts(Tb=bit period,Ts=sampling period)
# if a carrier is used, use L = Fs/Fc, where Fs >> 2xFc
Fc=800 # carrier frequency
Fs=L*Fc # sampling frequency

BER_suboptimum = np.zeros(len(EbN0dB)) # BER measures
BER_optimum = np.zeros(len(EbN0dB))

#----------------Transmitter--------------------
ak = np.random.randint(2, size=N) # uniform random symbols from 0's and 1's
bk = lfilter([1.0],[1.0,-1.0],ak) # IIR filter for differential encoding
bk = bk%2 #XOR operation is equivalent to modulo-2
[s_bb,t]= bpsk_mod(bk,L) # BPSK modulation(waveform) - baseband
s = s_bb*np.cos(2*np.pi*Fc*t/Fs).astype(complex) # DBPSK with carrier

for i,EbN0 in enumerate(EbN0dB):
    # Compute and add AWGN noise
    r = awgn(s,EbN0,L) # refer Chapter section 4.1

    #----------suboptimum receiver--------------
    p=np.real(r)*np.cos(2*np.pi*Fc*t/Fs) # demodulate to baseband using BPF
    w0= np.hstack((p,np.zeros(L))) # append L samples on one arm for equal lengths
    w1= np.hstack((np.zeros(L),p)) # delay the other arm by Tb (L samples)
    w = w0*w1 # multiplier
    z = np.convolve(w,np.ones(L)) #integrator from kTb to (K+1)Tb (L samples)
    u =  z[L-1:-1-L:L] # sampler t=kTb
    ak_hat = (u<0) #decision
    BER_suboptimum[i] = np.sum(ak!=ak_hat)/N #BER for suboptimum receiver

    #----------optimum receiver--------------
    p=np.real(r)*np.cos(2*np.pi*Fc*t/Fs); # multiply I arm by cos
    q=np.imag(r)*np.sin(2*np.pi*Fc*t/Fs) # multiply Q arm by sin
    x = np.convolve(p,np.ones(L)) # integrate I-arm by Tb duration (L samples)
    y = np.convolve(q,np.ones(L)) # integrate Q-arm by Tb duration (L samples)
    xk = x[L-1:-1:L] # Sample every Lth sample
    yk = y[L-1:-1:L] # Sample every Lth sample
    w0 = xk[0:-2] # non delayed version on I-arm
    w1 = xk[1:-1] # 1 bit delay on I-arm
```

```
    z0 = yk[0:-2] # non delayed version on Q-arm
    z1 = yk[1:-1] # 1 bit delay on Q-arm
    u =w0*w1 + z0*z1 # decision statistic
    ak_hat=(u<0) # threshold detection
    BER_optimum[i] = np.sum(ak[1:-1]!=ak_hat)/N # BER for optimum receiver

#------Theoretical Bit/Symbol Error Rates-------------
EbN0lins = 10**(EbN0dB/10) # converting dB values to linear scale
theory_DBPSK_optimum = 0.5*np.exp(-EbN0lins)
theory_DBPSK_suboptimum = 0.5*np.exp(-0.76*EbN0lins)
theory_DBPSK_coherent=erfc(np.sqrt(EbN0lins))*(1-0.5*erfc(np.sqrt(EbN0lins)))
theory_BPSK_conventional = 0.5*erfc(np.sqrt(EbN0lins))

#-------------Plotting-------------------------
fig, ax = plt.subplots(nrows=1,ncols = 1)
ax.semilogy(EbN0dB,BER_suboptimum,'k*',label='DBPSK subopt (sim)')
ax.semilogy(EbN0dB,BER_optimum,'b*',label='DBPSK opt (sim)')
ax.semilogy(EbN0dB,theory_DBPSK_suboptimum,'m-',label='DBPSK subopt (theory)')
ax.semilogy(EbN0dB,theory_DBPSK_optimum,'r-',label='DBPSK opt (theory)')
ax.semilogy(EbN0dB,theory_DBPSK_coherent,'k-',label='coherent DEBPSK')
ax.semilogy(EbN0dB,theory_BPSK_conventional,'b-',label='coherent BPSK')
ax.set_title('Probability of D-BPSK over AWGN')
ax.set_xlabel('$E_b/N_0 (dB)$');ax.set_ylabel('$Probability of Bit Error - P_b$')
ax.legend();fig.show()
```

2.5 Quadrature Phase Shift Keying (QPSK)

QPSK is a form of phase modulation technique, in which two information bits (combined as one symbol) are modulated at once, selecting one of the four possible carrier phase shift states. The QPSK signal within a symbol duration T_{sym} is defined as

$$s(t) = A\,cos\,[2\pi f_c t + \theta_n] \quad, 0 \leq t \leq T_{sym},\, n = 1,2,3,4 \quad (2.13)$$

where the signal phase is given by

$$\theta_n = (2n-1)\frac{\pi}{4} \quad (2.14)$$

Therefore, the four possible initial signal phases are $\pi/4, 3\pi/4, 5\pi/4$ and $7\pi/4$ radians. Equation 2.13 can be re-written as

$$s(t) = A\,cos\theta_n\,cos(2\pi f_c t) - A\,sin\theta_n\,sin(2\pi f_c t) \quad (2.15)$$
$$= s_{ni}\phi_i(t) + s_{nq}\phi_q(t) \quad (2.16)$$

The above expression indicates the use of two orthonormal basis functions: $\langle \phi_i(t), \phi_q(t) \rangle$ together with the inphase and quadrature signaling points: $\langle s_{ni}, s_{nq} \rangle$. Therefore, on a two dimensional co-ordinate system with the axes set to $\phi_i(t)$ and $\phi_q(t)$, the QPSK signal is represented by four constellation points dictated by the vectors $\langle s_{ni}, s_{nq} \rangle$ with $n = 1,2,3,4$.

2.5.1 QPSK transmitter

The QPSK transmitter, shown in Figure 2.10, is implemented as a python function qpsk_mod. In this implementation, a splitter separates the odd and even bits from the generated information bits. Each stream of odd bits (quadrature arm) and even bits (in-phase arm) are converted to NRZ format in a parallel manner.

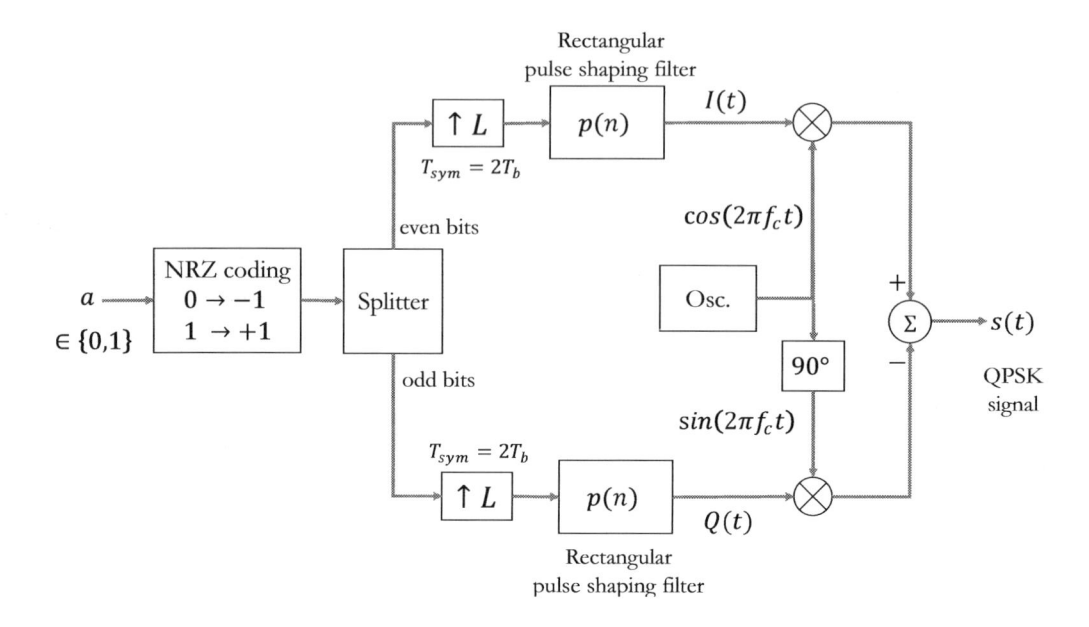

Fig. 2.10: Waveform simulation model for QPSK modulation

The timing diagram for BPSK and QPSK modulation is shown in Figure 2.11. For BPSK modulation the symbol duration for each bit is same as bit duration, but for QPSK the symbol duration is twice the bit duration: $T_{sym} = 2T_b$. Therefore, if the QPSK symbols were transmitted at same rate as BPSK, it is clear that QPSK sends twice as much data as BPSK does.

After oversampling and pulse shaping, it is intuitively clear that the signal on the I-arm and Q-arm are BPSK signals with symbol duration $2T_b$. The signal on the in-phase arm is then multiplied by $cos(2\pi f_c t)$ and the signal on the quadrature arm is multiplied by $-sin(2\pi f_c t)$. QPSK modulated signal is obtained by adding the signal from both in-phase and quadrature arms.

Note: The oversampling rate for the simulation is chosen as $L = 2f_s/f_c$, where f_c is the given carrier frequency and f_s is the sampling frequency satisfying Nyquist sampling theorem with respect to the carrier frequency ($f_s \geq fc$). This configuration gives integral number of carrier cycles for one symbol duration.

2.5 Quadrature Phase Shift Keying (QPSK)

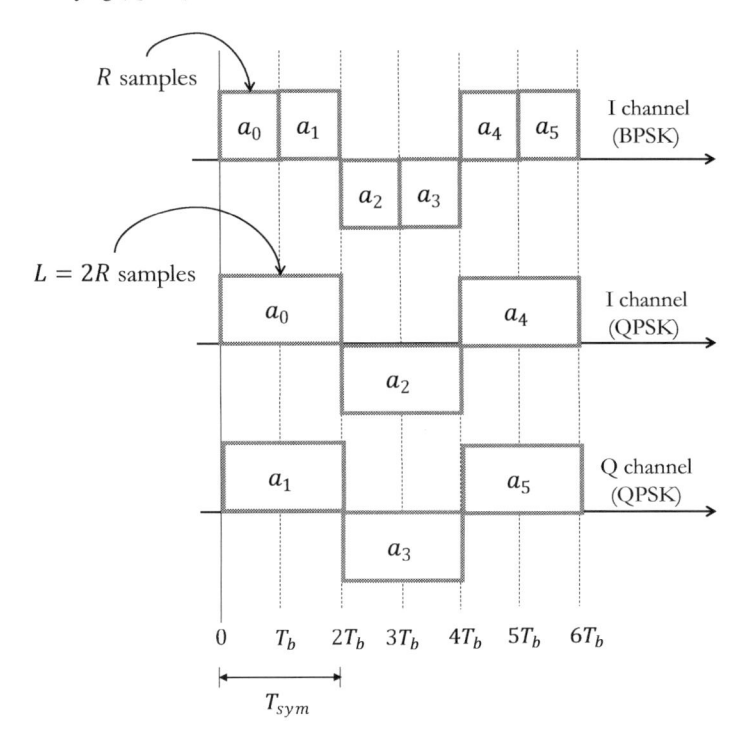

Fig. 2.11: Timing diagram for BPSK and QPSK modulations

Program 22: DigiCommPy\passband_modulations.py: QPSK modulator

```python
def qpsk_mod(a, fc, OF, enable_plot = False):
    """
    Modulate an incoming binary stream using conventional QPSK
    Parameters:
        a : input binary data stream (0's and 1's) to modulate
        fc : carrier frequency in Hertz
        OF : oversampling factor - at least 4 is better
        enable_plot : True = plot transmitter waveforms (default False)
    Returns:
        result : Dictionary containing the following keyword entries:
          s(t) : QPSK modulated signal vector with carrier i.e, s(t)
          I(t) : baseband I channel waveform (no carrier)
          Q(t) : baseband Q channel waveform (no carrier)
          t : time base for the carrier modulated signal
    """
    L = 2*OF # samples in each symbol (QPSK has 2 bits in each symbol)
    I = a[0::2];Q = a[1::2] #even and odd bit streams
    # even/odd streams at 1/2Tb baud
    
    from scipy.signal import upfirdn #NRZ encoder
    I = upfirdn(h=[1]*L, x=2*I-1, up = L)
    Q = upfirdn(h=[1]*L, x=2*Q-1, up = L)
```

```
        fs = OF*fc # sampling frequency
        t=np.arange(0,len(I)/fs,1/fs)   #time base

        I_t = I*np.cos(2*np.pi*fc*t);Q_t = -Q*np.sin(2*np.pi*fc*t)
        s_t = I_t + Q_t # QPSK modulated baseband signal

        if enable_plot:
            fig = plt.figure(constrained_layout=True)

            from matplotlib.gridspec import GridSpec
            gs = GridSpec(3, 2, figure=fig)
            ax1 = fig.add_subplot(gs[0, 0])
            ax2 = fig.add_subplot(gs[0, 1])
            ax3 = fig.add_subplot(gs[1, 0])
            ax4 = fig.add_subplot(gs[1, 1])
            ax5 = fig.add_subplot(gs[-1,:])

            # show first few symbols of I(t), Q(t)
            ax1.plot(t,I)
            ax2.plot(t,Q)
            ax3.plot(t,I_t,'r')
            ax4.plot(t,Q_t,'r')

            ax1.set_title('I(t)')
            ax2.set_title('Q(t)')
            ax3.set_title('$I(t) cos(2 \pi f_c t)$')
            ax4.set_title('$Q(t) sin(2 \pi f_c t)$')

            ax1.set_xlim(0,20*L/fs);ax2.set_xlim(0,20*L/fs)
            ax3.set_xlim(0,20*L/fs);ax4.set_xlim(0,20*L/fs)
            ax5.plot(t,s_t);ax5.set_xlim(0,20*L/fs);fig.show()
            ax5.set_title('$s(t) = I(t) cos(2 \pi f_c t) - Q(t) sin(2 \pi f_c t)$')
        result = dict()
        result['s(t)'] =s_t;result['I(t)'] = I;result['Q(t)'] = Q;result['t'] = t
        return result
```

2.5.2 QPSK receiver

Due to its special relationship with BPSK, the QPSK receiver takes the simplest form as shown in Figure 2.12. In this implementation, the I-channel and Q-channel signals are individually demodulated in the same way as that of BPSK demodulation. After demodulation, the I-channel bits and Q-channel sequences are combined into a single sequence. The function qpsk_demod implements a QPSK demodulator as per Figure 2.12

2.5 Quadrature Phase Shift Keying (QPSK)

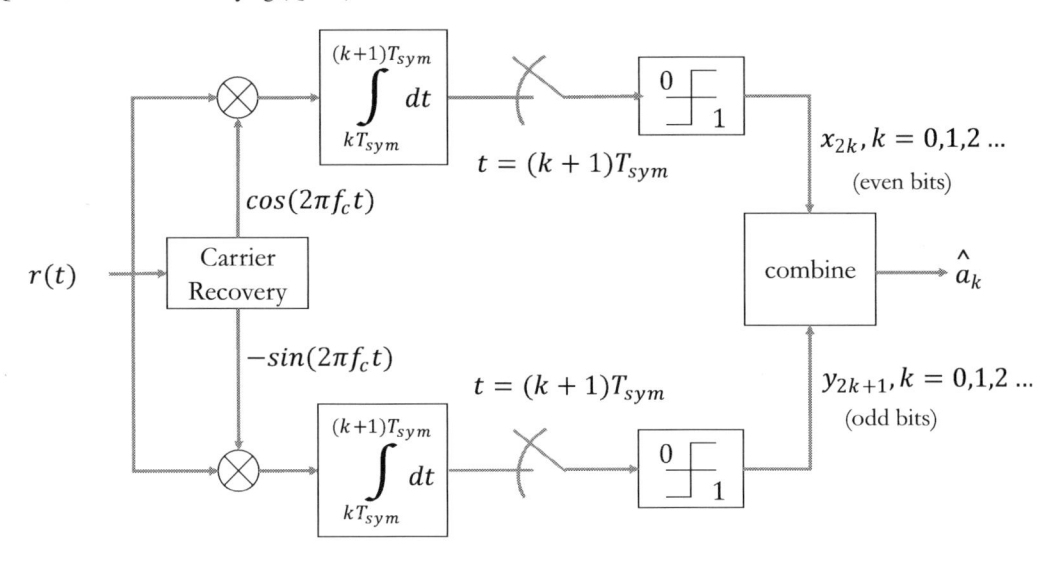

Fig. 2.12: Waveform simulation model for QPSK demodulation

Program 23: DigiCommPy\passband_modulations.py: QPSK demodulator

```python
def qpsk_demod(r,fc,OF,enable_plot=False):
    """
    Demodulate a conventional QPSK signal
    Parameters:
        r : received signal at the receiver front end
        fc : carrier frequency (Hz)
        OF : oversampling factor (at least 4 is better)
        enable_plot : True = plot receiver waveforms (default False)
    Returns:
        a_hat - detected binary stream
    """
    fs = OF*fc # sampling frequency
    L = 2*OF # number of samples in 2Tb duration
    t=np.arange(0,len(r)/fs,1/fs) # time base
    x=r*np.cos(2*np.pi*fc*t) # I arm
    y=-r*np.sin(2*np.pi*fc*t) # Q arm
    x = np.convolve(x,np.ones(L)) # integrate for L (Tsym=2*Tb) duration
    y = np.convolve(y,np.ones(L)) #integrate for L (Tsym=2*Tb) duration

    x = x[L-1::L] # I arm - sample at every symbol instant Tsym
    y = y[L-1::L] # Q arm - sample at every symbol instant Tsym
    a_hat = np.zeros(2*len(x))
    a_hat[0::2] = (x>0) # even bits
    a_hat[1::2] = (y>0) # odd bits

    if enable_plot:
        fig, axs = plt.subplots(1, 1)
```

```
        axs.plot(x[0:200],y[0:200],'o');fig.show()
    return a_hat
```

2.5.3 *Performance simulation over AWGN*

The complete waveform simulation for the aforementioned QPSK modulation and demodulation is given next. The simulation involves, generating random message bits, modulating them using QPSK modulation, addition of AWGN channel noise corresponding to the given signal-to-noise ratio and demodulating the noisy signal using a coherent QPSK receiver. The waveforms at the various stages of the modulator are shown in the Figure 2.13. The ideal-constellation at the transmitter and the constellation at the receiver that is affected by channel noise (shown for $E_b/N_0 = 10\ dB$) are shown in Figure 2.28 respectively.

Program 24: DigiCommPy\chapter_2\qpsk.py: Waveform simulation of performance of QPSK

```python
#Execute in Python3: exec(open("chapter_2/perf_qpsk.py").read())
import numpy as np #for numerical computing
import matplotlib.pyplot as plt #for plotting functions
from DigiCommPy.passband_modulations import qpsk_mod,qpsk_demod
from DigiCommPy.channels import awgn
from scipy.special import erfc

N=100000 # Number of symbols to transmit
EbN0dB = np.arange(start=-4,stop = 11,step = 2) # Eb/N0 range in dB for simulation
fc=100 # carrier frequency in Hertz
OF =8 # oversampling factor, sampling frequency will be fs=OF*fc

BER = np.zeros(len(EbN0dB)) # For BER values for each Eb/N0

a = np.random.randint(2, size=N) # uniform random symbols from 0's and 1's
result = qpsk_mod(a,fc,OF,enable_plot=False) #QPSK modulation
s = result['s(t)'] # get values from returned dictionary

for i,EbN0 in enumerate(EbN0dB):
    # Compute and add AWGN noise
    r = awgn(s,EbN0,OF) # refer Chapter section 4.1
    a_hat = qpsk_demod(r,fc,OF) # QPSK demodulation
    BER[i] = np.sum(a!=a_hat)/N # Bit Error Rate Computation

#------Theoretical Bit Error Rate-------------
theoreticalBER = 0.5*erfc(np.sqrt(10**(EbN0dB/10)))
#-------------Plot performance curve------------------
fig, axs = plt.subplots(nrows=1,ncols = 1)
axs.semilogy(EbN0dB,BER,'k*',label='Simulated')
axs.semilogy(EbN0dB,theoreticalBER,'r-',label='Theoretical')
axs.set_title('Probability of Bit Error for QPSK modulation');
axs.set_xlabel(r'$E_b/N_0$ (dB)')
axs.set_ylabel(r'Probability of Bit Error - $P_b$');
axs.legend();fig.show()
```

2.6 Offset QPSK (O-QPSK)

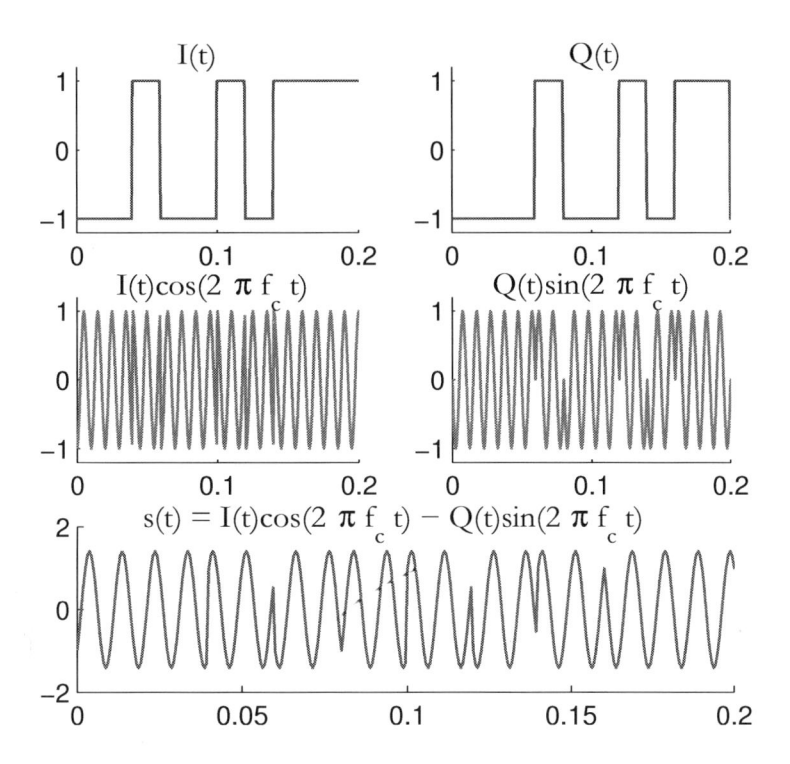

Fig. 2.13: Simulated QPSK waveforms at the transmitter side

The performance simulation for the QPSK transmitter-receiver combination was also coded in the code given above and the resulting bit-error rate performance curve will be same as that of conventional BPSK (Figure 2.6). A QPSK signal essentially combines two orthogonally modulated BPSK signals. Therefore, the resulting performance curves for QPSK - E_b/N_0 Vs. bits-in-error - will be same as that of conventional BPSK.

2.6 Offset QPSK (O-QPSK)

Offset QPSK is essentially same as QPSK, except that the orthogonal carrier signals on the I-channel and the Q-channel are staggered (one of them is delayed in time). OQPSK modulator and demodulator are shown in Figures 2.14 and 2.15 respectively. From the modulator, shown in Figure 2.14, it is clear that this scheme is same as QPSK except for an extra delay of half symbol period ($T_{sym}/2 = T_b$) in the Q-channel arm. OQPSK is also referred as staggered QPSK. Based on the modulator, an OQPSK signal can be written as

$$s(t) = I(t)\cos(2\pi f_c t) - Q\left(t - \frac{T_{sym}}{2}\right)\sin(2\pi f_c t) \quad (2.17)$$

The staggering concept in OQPSK is illustrated in Figure 2.16. Since the OQPSK differs from QPSK only in the time alignment of the orthogonal bit streams, both QPSK and OQPSK have the same power spectral density (refer the simulated PSD in Figure 2.29)and same error rate performance. However, these two modulations behave differently when passed through a bandlimited filter as encountered in certain applications like satellite communications. This difference in the behavior is due to the nature of phase changes in the carrier for these two modulations.

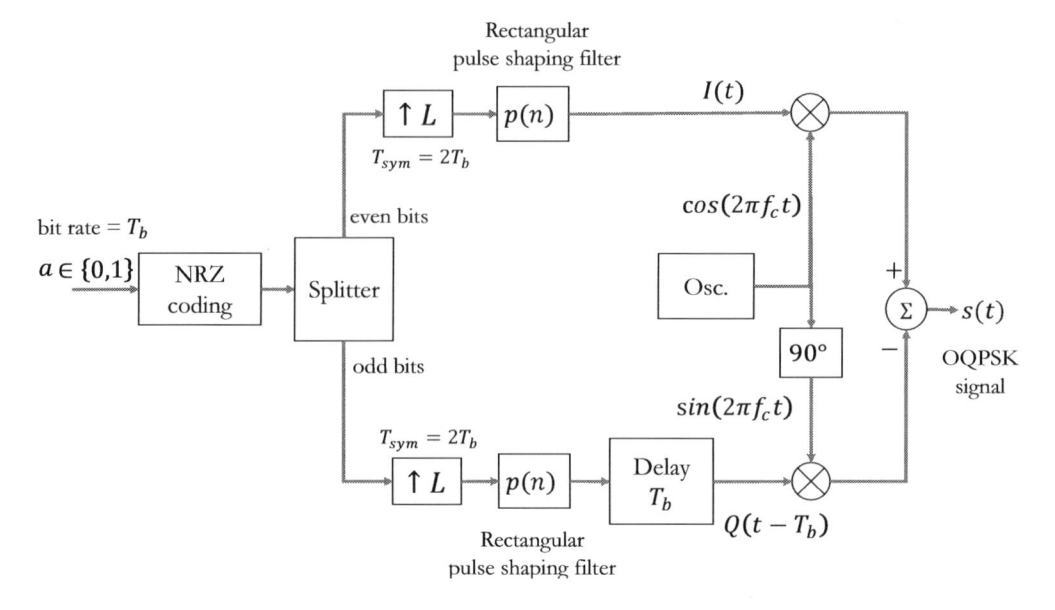

Fig. 2.14: OQPSK modulator

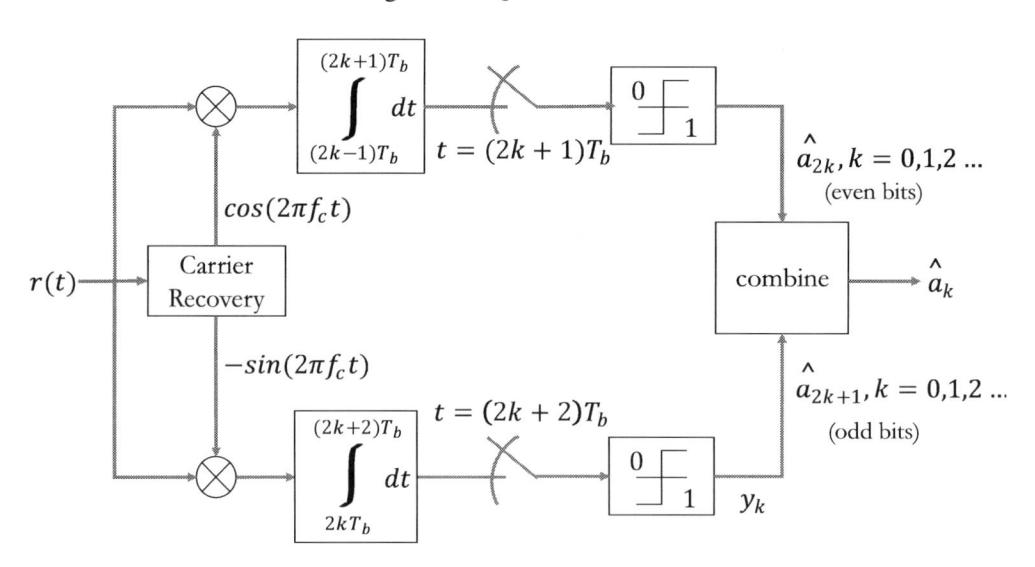

Fig. 2.15: OQPSK demodulator

In QPSK, due to coincident alignment of I and Q channels, the carrier phase changes at every symbol boundary, i.e, for every symbol duration ($T_{sym} = 2T_b$). If any one of the I-channel or Q-channel component changes sign, a phase shift of $\pm 90°$ occurs. If both the components change in sign, it results in a phase shift of $180°$. Phase shifts of $180°$ in QPSK degrades its constant envelop property and it can cause deleterious effects like out-of-band radiation when passed through bandlimited filters. On the otherhand, OQPSK signal does not suffer from these pitfalls.

In OQPSK, the orthogonal components cannot change states at the same time, this is because the components change state only at the middle of the symbol periods (due to the half symbol offset in the Q-channel). This eliminates $180°$ phase shifts all together and the phase changes are limited to $0°$ or $\pm 90°$ every T_b seconds.

2.6 Offset QPSK (O-QPSK)

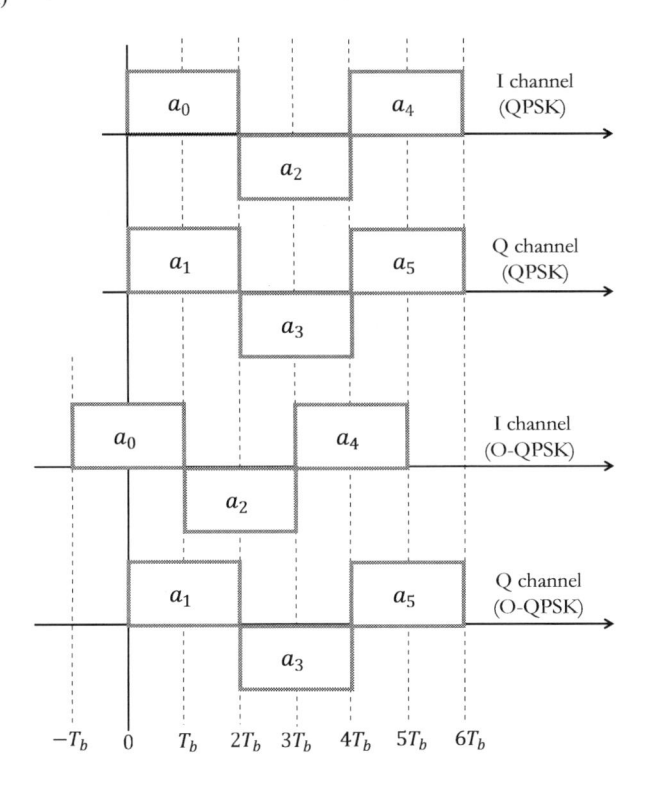

Fig. 2.16: Timing diagrams comparing QPSK and OQPSK modulations

Elimination of 180° phase shifts in OQPSK offers many advantages over QPSK. Unlike QPSK, the spectrum of OQPSK remains unchanged when bandlimited [4]. Additionally, OQPSK performs better than QPSK when subjected to phase jitters [5]. Further improvements to OQPSK can be obtained if the phase transitions are avoided altogether - as evident from continuous modulation schemes like Minimum Shift Keying (MSK) technique. The following functions implement the OQPSK modulator and demodulator given in Figures 2.14 and 2.15 respectively.

Program 25: DigiCommPy\passband_modulations.py: OQPSK modulator

```python
def oqpsk_mod(a,fc,OF,enable_plot=False):
    """
    Modulate an incoming binary stream using OQPSK
    Parameters:
        a : input binary data stream (0's and 1's) to modulate
        fc : carrier frequency in Hertz
        OF : oversampling factor - at least 4 is better
        enable_plot : True = plot transmitter waveforms (default False)
    Returns:
        result : Dictionary containing the following keyword entries:
          s(t) : QPSK modulated signal vector with carrier i.e, s(t)
          I(t) : baseband I channel waveform (no carrier)
          Q(t) : baseband Q channel waveform (no carrier)
          t : time base for the carrier modulated signal
    """
```

```python
L = 2*OF # samples in each symbol (QPSK has 2 bits in each symbol)
I = a[0::2];Q = a[1::2] #even and odd bit streams
# even/odd streams at 1/2Tb baud
from scipy.signal import upfirdn #NRZ encoder
I = upfirdn(h=[1]*L, x=2*I-1, up = L)
Q = upfirdn(h=[1]*L, x=2*Q-1, up = L)

I = np.hstack((I,np.zeros(L//2))) # padding at end
Q = np.hstack((np.zeros(L//2),Q)) # padding at start

fs = OF*fc # sampling frequency
t=np.arange(0,len(I)/fs,1/fs)  #time base
I_t = I*np.cos(2*np.pi*fc*t);Q_t = -Q*np.sin(2*np.pi*fc*t)
s = I_t + Q_t # QPSK modulated baseband signal

if enable_plot:
    fig = plt.figure(constrained_layout=True)

    from matplotlib.gridspec import GridSpec
    gs = GridSpec(3, 2, figure=fig)
    ax1 = fig.add_subplot(gs[0, 0]);ax2 = fig.add_subplot(gs[0, 1])
    ax3 = fig.add_subplot(gs[1, 0]);ax4 = fig.add_subplot(gs[1, 1])
    ax5 = fig.add_subplot(gs[-1,:])

    # show first few symbols of I(t), Q(t)
    ax1.plot(t,I);ax1.set_title('I(t)')
    ax2.plot(t,Q);ax2.set_title('Q(t)')
    ax3.plot(t,I_t,'r');ax3.set_title('$I(t) cos(2 \pi f_c t)$')
    ax4.plot(t,Q_t,'r');ax4.set_title('$Q(t) sin(2 \pi f_c t)$')
    ax1.set_xlim(0,20*L/fs);ax2.set_xlim(0,20*L/fs)
    ax3.set_xlim(0,20*L/fs);ax4.set_xlim(0,20*L/fs)
    ax5.plot(t,s);ax5.set_xlim(0,20*L/fs);fig.show()
    ax5.set_title('$s(t) = I(t) cos(2 \pi f_c t) - Q(t) sin(2 \pi f_c t)$')

    fig, axs = plt.subplots(1, 1)
    axs.plot(I,Q);fig.show()#constellation plot
result = dict()
result['s(t)'] =s;result['I(t)'] = I;result['Q(t)'] = Q;result['t'] = t
return result
```

Program 26: DigiCommPy\passband_modulations.py: OQPSK demodulator

```python
def oqpsk_demod(r,N,fc,OF,enable_plot=False):
    """
    Demodulate a OQPSK signal
    Parameters:
        r : received signal at the receiver front end
        N : Number of OQPSK symbols transmitted
        fc : carrier frequency (Hz)
        OF : oversampling factor (at least 4 is better)
```

2.6 Offset QPSK (O-QPSK)

```
            enable_plot : True = plot receiver waveforms (default False)
    Returns:
        a_hat - detected binary stream
    """
    fs = OF*fc # sampling frequency
    L = 2*OF # number of samples in 2Tb duration
    t=np.arange(0,(N+1)*OF/fs,1/fs) # time base
    x=r*np.cos(2*np.pi*fc*t) # I arm
    y=-r*np.sin(2*np.pi*fc*t) # Q arm
    x = np.convolve(x,np.ones(L)) # integrate for L (Tsym=2*Tb) duration
    y = np.convolve(y,np.ones(L)) #integrate for L (Tsym=2*Tb) duration

    x = x[L-1:-1-L:L] # I arm - sample at every symbol instant Tsym
    y = y[L+L//2-1:-1-L//2:L] # Q arm - sample at every symbol starting at L+L/2-1th
      ↪ sample
    a_hat = np.zeros(N)
    a_hat[0::2] = (x>0) # even bits
    a_hat[1::2] = (y>0) # odd bits

    if enable_plot:
        fig, axs = plt.subplots(1, 1)
        axs.plot(x[0:200],y[0:200],'o');fig.show()
    return a_hat
```

The wrapper code for simulating a complete communication chain with OQPSK modulator (with carrier), AWGN channel and OQPSK demodulator is given next. The code also includes performance simulation of the OQPSK communication system over an AWGN channel and the performance is essentially same as that of QPSK. The simulated timing waveforms at the transmitter is shown in Figure 2.17.

Program 27: DigiCommPy\chapter_2\oqpsk.py: Waveform simulation of performance of OQPSK

```
#Execute in Python3: exec(open("chapter_2/oqpsk.py").read())
import numpy as np #for numerical computing
import matplotlib.pyplot as plt #for plotting functions
from DigiCommPy.passband_modulations import oqpsk_mod,oqpsk_demod
from DigiCommPy.channels import awgn
from scipy.special import erfc

N=100000 # Number of symbols to transmit
EbN0dB = np.arange(start=-4,stop = 11,step = 2) # Eb/N0 range in dB for simulation
fc=100 # carrier frequency in Hertz
OF =8 # oversampling factor, sampling frequency will be fs=OF*fc

BER = np.zeros(len(EbN0dB)) # For BER values for each Eb/N0

a = np.random.randint(2, size=N) # uniform random symbols from 0's and 1's
result = oqpsk_mod(a,fc,OF,enable_plot=False) #QPSK modulation
s = result['s(t)'] # get values from returned dictionary
for i,EbN0 in enumerate(EbN0dB):
    # Compute and add AWGN noise
    r = awgn(s,EbN0,OF) # refer Chapter section 4.1
```

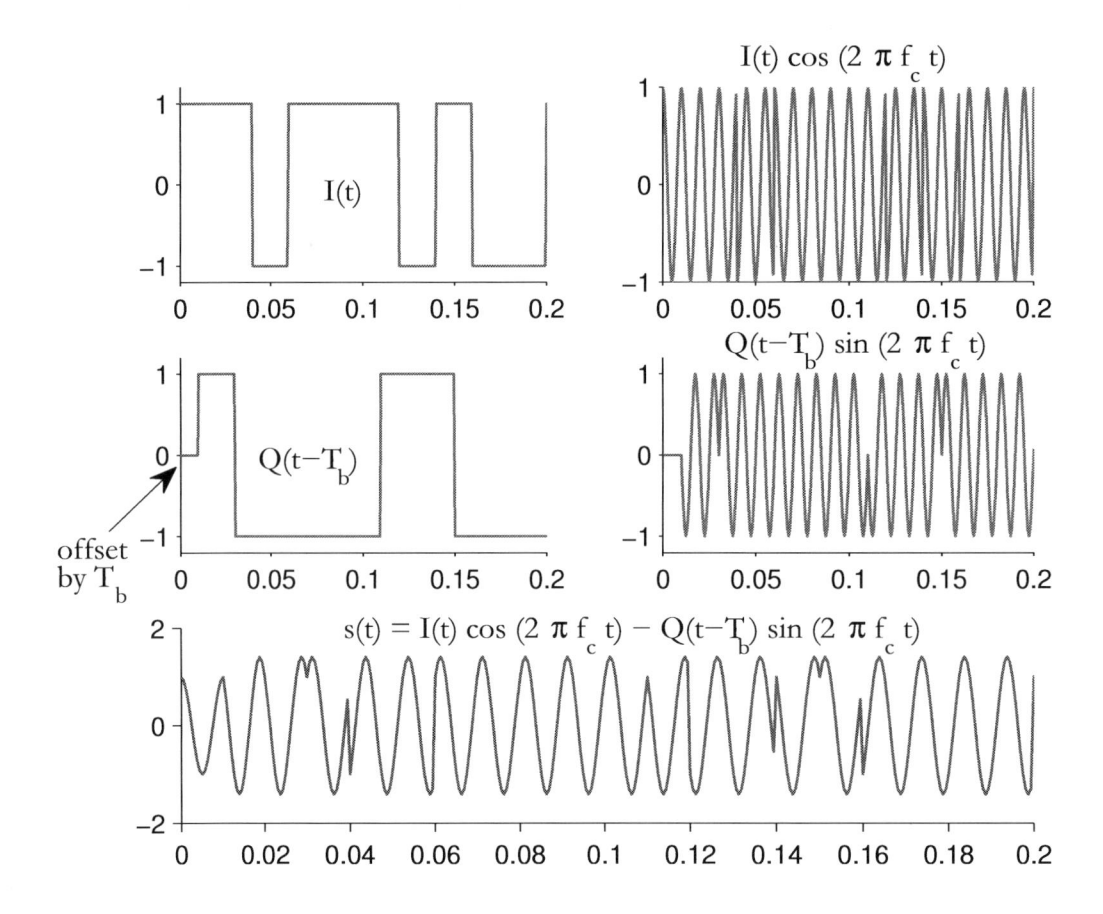

```
    a_hat = oqpsk_demod(r,N,fc,OF,enable_plot=False) # QPSK demodulation
    BER[i] = np.sum(a!=a_hat)/N # Bit Error Rate Computation
#------Theoretical Bit Error Rate-------------
theoreticalBER = 0.5*erfc(np.sqrt(10**(EbN0dB/10)))
#------------Plot performance curve---------------------
fig, axs = plt.subplots(nrows=1,ncols = 1)
axs.semilogy(EbN0dB,BER,'k*',label='Simulated')
axs.semilogy(EbN0dB,theoreticalBER,'r-',label='Theoretical')
axs.set_title('Probability of Bit Error for OQPSK');
axs.set_xlabel(r'$E_b/N_0$ (dB)')
axs.set_ylabel(r'Probability of Bit Error - $P_b$');
axs.legend();fig.show()
```

Fig. 2.17: Simulated OQPSK waveforms at the transmitter side

2.7 π/4-DQPSK

QPSK modulation has several variants, two such flavors among them are: $\pi/4$-QPSK and $\pi/4$-DQPSK.

In $\pi/4$-QPSK, the signaling points of the modulated signals are chosen from two QPSK constellations that are just shifted $\pi/4$ radians (45°) with respect to each other. Switching between the two constellations every successive bit ensures that the phase changes are confined to odd multiples of 45°. Therefore, phase transitions of ±90° and 180° are eliminated.

$\pi/4$-QPSK preserves the constant envelope property better than QPSK and OQPSK. Unlike QPSK and OQPSK schemes, $\pi/4$-QPSK can be differentially encoded therefore enabling the use of both coherent and non-coherent demodulation techniques. Choice of non-coherent demodulation results in simpler receiver design. Differentially encoded $\pi/4$-QPSK is called $\pi/4$-*DQPSK* which is the subject of this section.

Modulator

The modulator for $\pi/4$-DQPSK is shown in Figure 2.18. The differential encoding for $\pi/4$-DQPSK is defined by the following encoding rules.

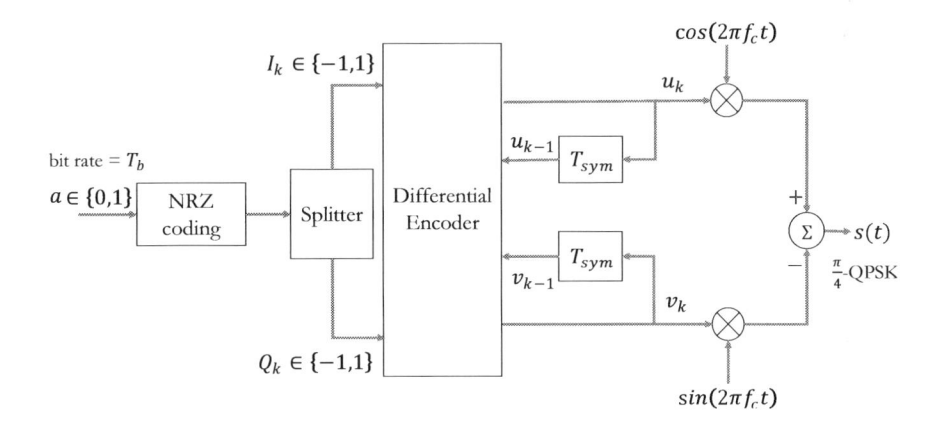

Fig. 2.18: $\pi/4$-DQPSK modulator

Let $\langle I(t), Q(t) \rangle$ and $\langle u(t), v(t) \rangle$ be the unencoded and differentially encoded versions of I-channel and Q-channel waveform sequences. From reference [3], the differentially encoded $\pi/4$-DQPSK signals are obtained as

$$u_k = u_{k-1} \cos\Delta\theta_k - v_{k-1} \sin\Delta\theta_k$$
$$v_k = u_{k-1} \sin\Delta\theta_k + v_{k-1} \cos\Delta\theta_k \tag{2.18}$$

where, $\Delta\theta_k$ defines the phase difference determined by the input data as given in Table 2.1. The initial values for the very first encoded samples are defined as $u_{-1} = 1$ and $v_{-1} = 0$. After applying the encoding rule given above, the differentially encoded I-channel and Q-channel bits are modulated in the same fashion as the conventional QPSK scheme.

The following Python function - `piBy4_dqpsk_diff_encoding`, implements the encoding rules for the differential encoder. The next shown function - `piBy4_dqpsk_mod`, implements the $\pi/4$-DQPSK modulator, as per Figure 2.18.

$I_k\ Q_k$	$\Delta\theta_k$	$cos\Delta\theta_k$	$sin\Delta\theta_k$
$-1\ -1$	$-3\pi/4$	$-1/\sqrt{2}$	$-1/\sqrt{2}$
$-1\ \ \ 1$	$3\pi/4$	$-1/\sqrt{2}$	$1/\sqrt{2}$
$\ \ 1\ -1$	$-\pi/4$	$1/\sqrt{2}$	$-1/\sqrt{2}$
$\ \ 1\ \ \ 1$	$\pi/4$	$1/\sqrt{2}$	$1/\sqrt{2}$

Table 2.1: Phase mapper for $\pi/4$-DQPSK

Program 28: DigiCommPy\passband_modulations.py: Differential encoding for $\pi/4 - DQPSK$

```python
def piBy4_dqpsk_diff_encoding(a,enable_plot=False):
    """
    Phase Mapper for pi/4-DQPSK modulation
    Parameters:
        a : input stream of binary bits
    Returns:
        (u,v): tuple, where
            u : differentially coded I-channel bits
            v : differentially coded Q-channel bits
    """
    from numpy import pi, cos, sin
    if len(a)%2: raise ValueError('Length of binary stream must be even')
    I = a[0::2] # odd bit stream
    Q = a[1::2] # even bit stream
    # club 2-bits to form a symbol and use it as index for dTheta table
    m = 2*I+Q
    dTheta = np.array([-3*pi/4, 3*pi/4, -pi/4, pi/4]) #LUT for pi/4-DQPSK
    u = np.zeros(len(m)+1);v = np.zeros(len(m)+1)
    u[0]=1; v[0]=0 # initial conditions for uk and vk
    for k in range(0,len(m)):
        u[k+1] = u[k] * cos(dTheta[m[k]]) - v[k] * sin(dTheta[m[k]])
        v[k+1] = u[k] * sin(dTheta[m[k]]) + v[k] * cos(dTheta[m[k]])
    if enable_plot:#constellation plot
        fig, axs = plt.subplots(1, 1)
        axs.plot(u,v,'o');
        axs.set_title('Constellation');fig.show()
    return (u,v)
```

Program 29: DigiCommPy\passband_modulations.py: $\pi/4 - DQPSK$ modulator

```python
def piBy4_dqpsk_mod(a,fc,OF,enable_plot = False):
    """
    Modulate a binary stream using pi/4 DQPSK
    Parameters:
        a : input binary data stream (0's and 1's) to modulate
        fc : carrier frequency in Hertz
        OF : oversampling factor
    Returns:
        result : Dictionary containing the following keyword entries:
```

2.7 π/4-DQPSK

```
            s(t) : pi/4 QPSK modulated signal vector with carrier
            U(t) : differentially coded I-channel waveform (no carrier)
            V(t) : differentially coded Q-channel waveform (no carrier)
            t: time base
    """
    (u,v)=piBy4_dqpsk_diff_encoding(a) # Differential Encoding for pi/4 QPSK
    #Waveform formation (similar to conventional QPSK)
    L = 2*OF # number of samples in each symbol (QPSK has 2 bits/symbol)
    U = np.tile(u, (L,1)).flatten('F')# odd bit stream at 1/2Tb baud
    V = np.tile(v, (L,1)).flatten('F')# even bit stream at 1/2Tb baud

    fs = OF*fc # sampling frequency
    t=np.arange(0, len(U)/fs,1/fs) #time base
    U_t = U*np.cos(2*np.pi*fc*t)
    V_t = -V*np.sin(2*np.pi*fc*t)
    s_t = U_t + V_t

    if enable_plot:
        fig = plt.figure(constrained_layout=True)

        from matplotlib.gridspec import GridSpec
        gs = GridSpec(3, 2, figure=fig)
        ax1 = fig.add_subplot(gs[0, 0]);ax2 = fig.add_subplot(gs[0, 1])
        ax3 = fig.add_subplot(gs[1, 0]);ax4 = fig.add_subplot(gs[1, 1])
        ax5 = fig.add_subplot(gs[-1,:])
        ax1.plot(t,U);ax2.plot(t,V)
        ax3.plot(t,U_t,'r');ax4.plot(t,V_t,'r')
        ax5.plot(t,s_t) #QPSK waveform zoomed to first few symbols
        ax1.set_ylabel('U(t)-baseband');ax2.set_ylabel('V(t)-baseband')
        ax3.set_ylabel('U(t)-with carrier');ax4.set_ylabel('V(t)-with carrier')
        ax5.set_ylabel('s(t)');ax5.set_xlim([0,10*L/fs])
        ax1.set_xlim([0,10*L/fs]);ax2.set_xlim([0,10*L/fs])
        ax3.set_xlim([0,10*L/fs]);ax4.set_xlim([0,10*L/fs])
        fig.show()

    result = dict()
    result['s(t)'] =s_t;result['U(t)'] = U;result['V(t)'] = V;result['t'] = t
    return result
```

Demodulator

From the equations for π/4-DQPSK modulation, we can see that the information is encoded as phase difference $\Delta\theta_k$. Therefore, in addition to coherent detection, non-coherent detection techniques can also be used for demodulating the π/4-DQPSK signal. Of the several demodulation techniques available for demodulating the π/4-DQPSK signal, the baseband differential detection is used here for simulation. Refer [3] for details on the rest of the demodulator techniques.

The demodulator for baseband differential detection technique is presented in Figure 2.19. Let w_k and z_k be the baseband versions of differentially encoded I-channel and Q-channel bits at the receiver (they are essentially the output of the symbol rate samplers shown in the Figure 2.19). The decoding rules for obtaining the corresponding differentially decoded versions are given by

$$x_k = w_k w_{k-1} + z_k z_{k-1}$$
$$y_k = z_k w_{k-1} - w_k z_{k-1}$$

(2.19)

The I-channel and Q-channel bits are then decided as

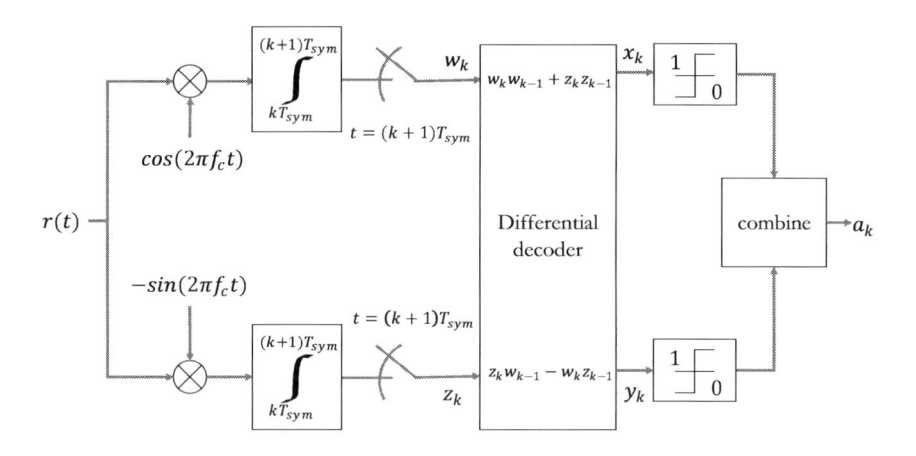

$$\hat{I}_k \triangleq x_k \underset{0}{\overset{1}{\gtrless}} 0 \tag{2.20}$$

$$\hat{Q}_k \triangleq y_k \underset{0}{\overset{1}{\gtrless}} 0 \tag{2.21}$$

Fig. 2.19: Baseband differential demodulator for $\pi/4$-DQPSK

The Python function (piBy4_dqpsk_diff_decoding) implementing the above mentioned equations for the differential decoder is given next. The function implementing the $\pi/4$-DQPSK demodulator as per Figure 2.19, is also given.

Program 30: DigiCommPy\passband_modulations.py: $\pi/4 - DQPSK$ differential decoding detection

```python
def piBy4_dqpsk_diff_decoding(w,z):
    """
    Phase Mapper for pi/4-DQPSK modulation
    Parameters:
        w - differentially coded I-channel bits at the receiver
        z - differentially coded Q-channel bits at the receiver
    Returns:
        a_hat - binary bit stream after differential decoding
    """
    if len(w)!=len(z): raise ValueError('Length mismatch between w and z')
```

2.7 $\pi/4$-DQPSK

```python
    x = np.zeros(len(w)-1);y = np.zeros(len(w)-1);
    
    for k in range(0,len(w)-1):
        x[k] = w[k+1]*w[k] + z[k+1]*z[k]
        y[k] = z[k+1]*w[k] - w[k+1]*z[k]
    
    a_hat = np.zeros(2*len(x))
    a_hat[0::2] = (x > 0) # odd bits
    a_hat[1::2] = (y > 0) # even bits
    return a_hat
```

Program 31: DigiCommPy\passband_modulations.py: $\pi/4 - DQPSK$ demodulator

```python
def piBy4_dqpsk_demod(r,fc,OF,enable_plot=False):
    """
    Differential coherent demodulation of pi/4-DQPSK
    Parameters:
        r : received signal at the receiver front end
        fc : carrier frequency in Hertz
        OF : oversampling factor (multiples of fc) - at least 4 is better
    Returns:
        a_cap :  detected binary stream
    """
    fs = OF*fc # sampling frequency
    L = 2*OF # samples in 2Tb duration
    t=np.arange(0, len(r)/fs,1/fs)
    w=r*np.cos(2*np.pi*fc*t) # I arm
    z=-r*np.sin(2*np.pi*fc*t) # Q arm
    w = np.convolve(w,np.ones(L)) # integrate for L (Tsym=2*Tb) duration
    z = np.convolve(z,np.ones(L)) # integrate for L (Tsym=2*Tb) duration
    w = w[L-1::L] # I arm - sample at every symbol instant Tsym
    z = z[L-1::L] # Q arm - sample at every symbol instant Tsym
    a_cap = piBy4_dqpsk_diff_decoding(w,z)
    
    if enable_plot:#constellation plot
        fig, axs = plt.subplots(1, 1)
        axs.plot(w,z,'o')
        axs.set_title('Constellation');fig.show()
    return a_cap
```

Performance over AWGN channel

The wrapper code for simulating a complete communication chain with $\pi/4$-DQPSK modulator, AWGN channel and $\pi/4$-DQPSK demodulator is given next. The simulated waveforms at the transmitter are shown in Figure 2.20(a). The code also includes performance simulation of the $\pi/4$-DQPSK communication system over a range of E_b/N_0 values. The simulated performance curve is plotted in Figure 2.20(b) along with the theoretical bit error probability of differentially coherently demodulated $\pi/4$-DQPSK, which is given by

$$P_b = Q\left(\sqrt{\frac{4E_b}{N_0}}\, sin\frac{\pi}{4\sqrt{2}}\right)$$
$$= \frac{1}{2}\, erfc\left(\frac{1}{\sqrt{2}}\sqrt{\frac{4E_b}{N_0}}\, sin\frac{\pi}{4\sqrt{2}}\right) \quad (2.22)$$

Program 32: DigiCommPy\chapter_2\piby4_dqpsk.py: $\pi/4 - DQPSK$ performance simulation

```python
#Execute in Python3: exec(open("chapter_2/piby4_dqpsk.py").read())
import numpy as np #for numerical computing
import matplotlib.pyplot as plt #for plotting functions
from DigiCommPy.passband_modulations import piBy4_dqpsk_mod,piBy4_dqpsk_demod
from DigiCommPy.channels import awgn
from scipy.special import erfc

N=1000000 # Number of symbols to transmit
EbN0dB = np.arange(start=-4,stop = 11,step = 2) # Eb/N0 range in dB for simulation
fc=100 # carrier frequency in Hertz
OF =8 # oversampling factor, sampling frequency will be fs=OF*fc

BER = np.zeros(len(EbN0dB)) # For BER values for each Eb/N0

a = np.random.randint(2, size=N) # uniform random symbols from 0's and 1's
result = piBy4_dqpsk_mod(a,fc,OF,enable_plot=False)# dqpsk modulation
s = result['s(t)'] # get values from returned dictionary

for i,EbN0 in enumerate(EbN0dB):
    # Compute and add AWGN noise
    r = awgn(s,EbN0,OF) # refer Chapter section 4.1
    a_hat = piBy4_dqpsk_demod(r,fc,OF,enable_plot=False)
    BER[i] = np.sum(a!=a_hat)/N # Bit Error Rate Computation

#------Theoretical Bit Error Rate-------------
x = np.sqrt(4*10**(EbN0dB/10))*np.sin(np.pi/(4*np.sqrt(2)))
theoreticalBER = 0.5*erfc(x/np.sqrt(2))

#-------------Plot performance curve----------------------
fig, axs = plt.subplots(nrows=1,ncols = 1)
axs.semilogy(EbN0dB,BER,'k*',label='Simulated')
axs.semilogy(EbN0dB,theoreticalBER,'r-',label='Theoretical')
axs.set_title('Probability of Bit Error for $\pi/4$-DQPSK');
axs.set_xlabel(r'$E_b/N_0$ (dB)')
axs.set_ylabel(r'Probability of Bit Error - $P_b$');
axs.legend();fig.show()
```

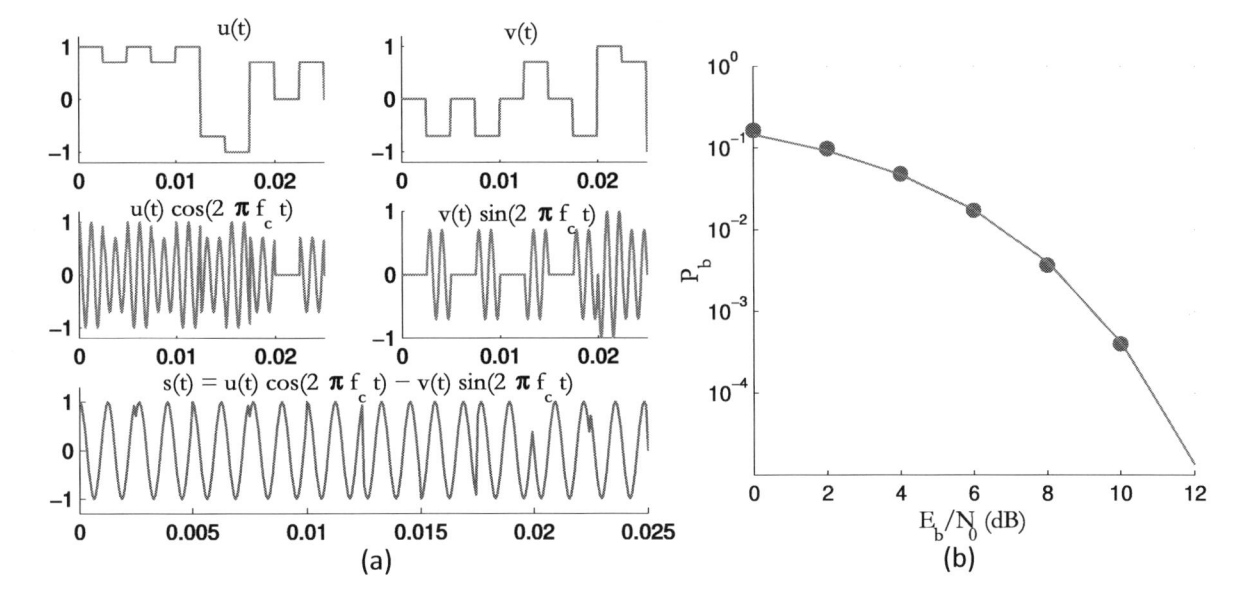

Fig. 2.20: (a) Simulated $\pi/4$-DQPSK waveforms at the transmitter side, (b) Performance of differential coherently demodulated $\pi/4$-DQPSK

2.8 Continuous Phase Modulation (CPM)

2.8.1 Motivation behind CPM

Modulation schemes like conventional PSK or FSK introduce abrupt phase discontinuities in the transmitted signal. Abrupt phase discontinuities produce sidelobes and could cause interference on the nearby channels. It is highly desirable that the chosen modulation scheme for a radio communication, should produce a spectrally efficient signal (minimized sidelobe effects) so that more users can be served in the allocated frequency band. Additionally, the phase discontinuities can also introduce undesirable frequency sidelobes when passed through power-efficient non-linear amplifiers (Class C).

In the case of conventional PSK or FSK modulation, the carrier phase abruptly changes at the start of every transmitted symbol. If the change of carrier phase can be made smooth from one symbol to another symbol, the sidelobe levels can be reduced.

Offset QPSK (OQPSK) modulation, discussed in section 2.6, is a special case of QPSK modulation that limits the range of phase variations in the transmitted signal. Therefore, some improvement in the sidelobe levels can be achieved with OQPSK. Further improvement can be obtained with Continuous Phase Modulation (CPM) schemes where the carrier phase smoothly varies from one symbol to another symbol in a continuous manner. It is important to note that OQPSK is not a continuous phase modulation scheme. It simply restricts the range of phase variations and does not make the phase variations continuous.

2.8.2 Continuous Phase Frequency Shift Keying (CPFSK) modulation

In CPM, the phase of the carrier is gradually changed from one symbol to another symbol. The instantaneous phase depends on the phase of previous symbols. Therefore, CPM is classified as *modulation with memory*. The modulation and demodulation of CPM are more complicated by the fact that the phase of one symbol depends on the phase of the previous symbols. The phase trajectories of CPM modulated signals are best exploited by *maximum likelihood sequence estimation* (MLSE) technique implemented by Viterbi algorithm. The demodulation techniques for CPM are more complex than the techniques used for memoryless modulation schemes like BPSK,QPSK,QAM,FSK etc.

Continuous Phase Frequency Shift Keying (CPFSK) is a variation of *frequency shift keying* (FSK) technique, that eliminates phase discontinuities. In conventional binary FSK modulation, the instantaneous carrier frequency is switched between one of the two frequencies. Due to the abrupt switching between the two frequencies, the phase is not continuous in conventional FSK. In binary CPFSK, the smooth phase continuity is ensured by defining the transmitter signal as

$$s(t) = \sqrt{\frac{2E_b}{T_b}} cos\left[2\pi f_c t + \theta(t)\right] \quad (2.23)$$

where, E_b is the average energy per bit, T_b is the bit period, f_c is the *base carrier frequency* and $\theta(t)$ is the *phase evolution* whose derivative gives rise to CPFSK's instantaneous angular frequency shift.

At any time instant, the phase evolution can be given as

$$\theta(t) = \theta(0) + \frac{\pi h}{T_b} \int_0^t b(t) dt \quad (2.24)$$

where $\theta(0)$ being the accumulated phase history till $t = 0$, the factor h is called *modulation index* which is a measure of frequency deviation ($h = 1$ corresponds to binary CPFSK and $h = 0.5$ corresponds to *Minimum Shift Keying* (MSK) modulation), $b(t) \in (\pm 1V)$ is the waveform that represents the binary information sequence a: such that the value $b = +1V$ represents $a = 0$ and the value $b = -1V$ represents $a = 1$.

The following snippet of code simulates the binary CPFSK modulation ($h = 1$) using equations 2.23, 2.24 and the resulting waveforms are plotted in Figure 2.21. The phase trajectory plotted in Figure 2.21-(b) shows ambiguity in the phase transitions ($+\pi$ phase transition is equivalent to $-\pi$ transition due to modulo 2π arithmetic). Due to this, the receiver will not be able to exploit the phase information in the binary CPFSK.

Program 33: DigiCommPy\chapter_2\cpfsk.py: Binary CPFSK modulation

```
import numpy as np
import matplotlib.pyplot as plt
from scipy.signal import lfilter

L = 50 # oversampling factor
Tb = 0.5 # bit period in seconds
fs = L/Tb # sampling frequency in Hertz
fc = 2/Tb # carrier frequency
N = 8 # number of bits to transmit
h = 1 # modulation index

b = 2*np.random.randint(2, size=N)-1 # random information sequence in +1/-1 format
b = np.tile(b, (L,1)).flatten('F')
b_integrated = lfilter([1.0],[1.0,-1.0],b)/fs #Integrate b using filter

theta= np.pi*h/Tb*b_integrated
t=np.arange(0, Tb*N, 1/fs) # time base

s = np.cos(2*np.pi*fc*t + theta) # CPFSK signal

fig, (ax1,ax2,ax3) = plt.subplots(3, 1)
```

2.8 Continuous Phase Modulation (CPM)

```
ax1.plot(t,b);ax1.set_xlabel('t');ax1.set_ylabel('b(t)')
ax2.plot(t,theta);ax2.set_xlabel('t');ax2.set_ylabel('$\theta(t)$')
ax3.plot(t,s);ax3.set_xlabel('t');ax3.set_ylabel('s(t)')
fig.show()
```

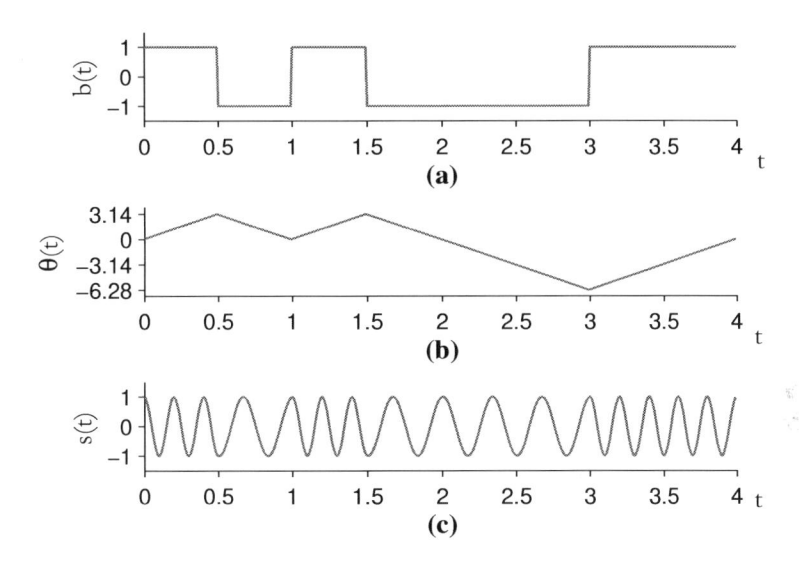

Fig. 2.21: (a) Information sequence, (b) phase evolution of CPFSK signal and (c) the CPFSK modulated signal

2.8.3 Minimum Shift Keying (MSK)

The minimum shift keying technique is a true CPM modulation technique. MSK modulation provides all the desired qualities loved by the communication engineers - it provides constant envelope, a very compact spectrum compared to QPSK and OQPSK, and a good error rate performance.

MSK can be viewed as a special case of binary CPFSK where the modulation index h in the equation 2.24 is set to 0.5. The same snippet of code given for CPFSK simulation is executed with h set to 0.5 and the results are plotted in Figure 2.22. The phase trajectory of MSK in Figure 2.22-(b), reveals that each information bit leads to different phase transitions on modulo-2π. The receiver can exploit these phase transitions without any ambiguity and it can provide better error rate performance. This is the main motivation behind the MSK technique.

Without loss of generality, assuming $\theta(0) = 0$ and setting $h = 0.5$ in equation 2.24, the equation 2.23 is modified to generate an MSK signal as

$$s(t) = \sqrt{\frac{2E_b}{T_b}} \cos\left[\theta(0) \pm \frac{\pi}{2T_b}t\right] \cos(2\pi f_c t) - \sqrt{\frac{2E_b}{T_b}} \sin\left[\theta(0) \pm \frac{\pi}{2T_b}t\right] \sin(2\pi f_c t)$$
$$= s_I(t)\cos(2\pi f_c t) - s_Q(t)\sin(2\pi f_c t) \quad , 0 \leq t \leq 2T_b \quad (2.25)$$

where the inphase component $s_I(t)$ and the quadrature component $s_Q(t)$ can be re-written as

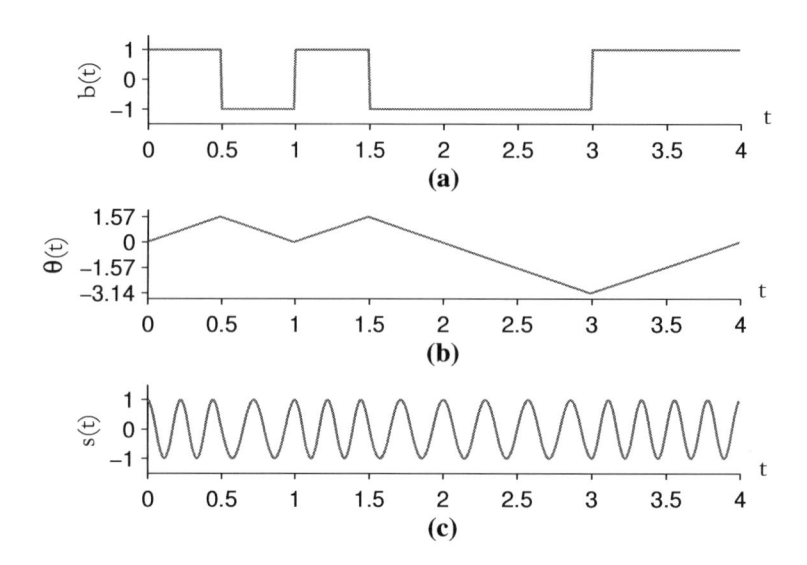

Fig. 2.22: (a) Information sequence, (b) phase evolution of MSK signal and (c) the MSK modulated signal

$$s_I(t) = \pm\sqrt{\frac{2E_b}{T_b}} \cos\left[\frac{\pi}{2T_b}t\right] \quad , \quad -T_b \leq t \leq T_b$$
$$s_Q(t) = \pm\sqrt{\frac{2E_b}{T_b}} \sin\left[\frac{\pi}{2T_b}t\right] \quad , \quad 0 \leq t \leq 2T_b \tag{2.26}$$

Therefore, rewriting equation 2.25, the MSK waveform is given by

$$s(t) = \sqrt{\frac{2}{T_b}} a_I(t) \cos\left[\frac{\pi}{2T_b}t\right] \cos(2\pi f_c t) - \sqrt{\frac{2}{T_b}} a_Q(t) \sin\left[\frac{\pi}{2T_b}t\right] \sin(2\pi f_c t) \tag{2.27}$$

where $a_I(t)$ and $a_Q(t)$ are random information sequences in the I-channel and Q-channel respectively.

Equivalence to OQPSK modulation

From equation 2.26, the inphase component $s_I(t)$ is interpreted as a half-cycle cosine function for the whole interval $(-T_b, T_b]$ and the quadrature component $s_Q(t)$ is interpreted as a half-cycle sine function for the interval $(0, 2T_b]$. Therefore, the half-cycle cosine and sine functions are offset from each other by T_b seconds. This offset relationship between the inphase and quadrature components is more similar to that of a OQPSK signal construct [6].

Figure 2.23 illustrates the similarities between the OQPSK and MSK signal construction. In OQPSK, the rectangular shaped inphase and quadrature components are offset by half symbol period ($T_{sym}/2 = T_b$ seconds). Whereas, in MSK modulation, the inphase and quadrature components are similarly offset by half symbol period ($T_{sym}/2 = T_b$ seconds) but they are additionally shaped by half-cycle cosine and sine functions.

2.8 Continuous Phase Modulation (CPM)   89

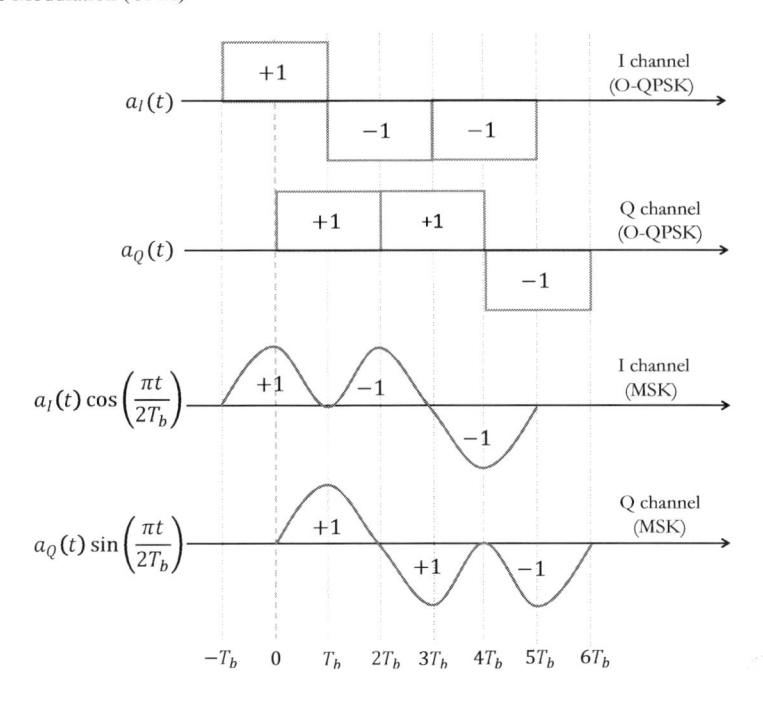

Fig. 2.23: Similarities between OQPSK and MSK waveforms

MSK modulator

Several forms of MSK signal generation/detection techniques exist and a good analysis can be found in reference [7]. As discussed above, MSK can be viewed as a special form of OQPSK with sinusoidal weighting. A practical MSK modulator that is more similar to the OQPSK modulator structure is given in Figure 2.24 and the corresponding Python function (msk_mod) is also given next. Here, the pulse shaping functions on the inphase and quadrature arms are implemented using identical low-pass filters having the following impulse response.

$$h(t) = \begin{cases} sin\left(\frac{\pi}{2T_b}t\right) & , 0 \leq t \leq 2T_b \\ 0 & , \text{otherwise} \end{cases} \quad (2.28)$$

Program 34: DigiCommPy\passband_modulations.py: MSK modulator

```
def msk_mod(a, fc, OF, enable_plot = False):
    """
    Modulate an incoming binary stream using MSK
    Parameters:
        a : input binary data stream (0's and 1's) to modulate
        fc : carrier frequency in Hertz
        OF : oversampling factor (at least 4 is better)
    Returns:
        result : Dictionary containing the following keyword entries:
          s(t) : MSK modulated signal with carrier
          sI(t) : baseband I channel waveform(no carrier)
          sQ(t) : baseband Q channel waveform(no carrier)
```

```
        t: time base
"""
ak = 2*a-1 # NRZ encoding 0-> -1, 1->+1
ai = ak[0::2]; aq = ak[1::2] # split even and odd bit streams
L = 2*OF # represents one symbol duration Tsym=2xTb

#upsample by L the bits streams in I and Q arms
from scipy.signal import upfirdn, lfilter
ai = upfirdn(h=[1], x=ai, up = L)
aq = upfirdn(h=[1], x=aq, up = L)

aq = np.pad(aq, (L//2,0), 'constant') # delay aq by Tb (delay by L/2)
ai = np.pad(ai, (0,L//2), 'constant') # padding at end to equal length of Q

#construct Low-pass filter and filter the I/Q samples through it
Fs = OF*fc;Ts = 1/Fs;Tb = OF*Ts
t = np.arange(0,2*Tb+Ts,Ts)
h = np.sin(np.pi*t/(2*Tb))# LPF filter
sI_t = lfilter(b = h, a = [1], x = ai) # baseband I-channel
sQ_t = lfilter(b = h, a = [1], x = aq) # baseband Q-channel

t=np.arange(0, Ts*len(sI_t), Ts) # for RF carrier
sIc_t = sI_t*np.cos(2*np.pi*fc*t) #with carrier
sQc_t = sQ_t*np.sin(2*np.pi*fc*t) #with carrier
s_t =  sIc_t - sQc_t# Bandpass MSK modulated signal

if enable_plot:
    fig, (ax1,ax2,ax3) = plt.subplots(3, 1)

    ax1.plot(t,sI_t);ax1.plot(t,sIc_t,'r')
    ax2.plot(t,sQ_t);ax2.plot(t,sQc_t,'r')
    ax3.plot(t,s_t,'--')
    ax1.set_ylabel('$s_I(t)$');ax2.set_ylabel('$s_Q(t)$')
    ax3.set_ylabel('s(t)')
    ax1.set_xlim([-Tb,20*Tb]);ax2.set_xlim([-Tb,20*Tb])
    ax3.set_xlim([-Tb,20*Tb])
    fig.show()

result = dict()
result['s(t)']=s_t;result['sI(t)']=sI_t;result['sQ(t)']=sQ_t;result['t']=t
return result
```

2.8 Continuous Phase Modulation (CPM)

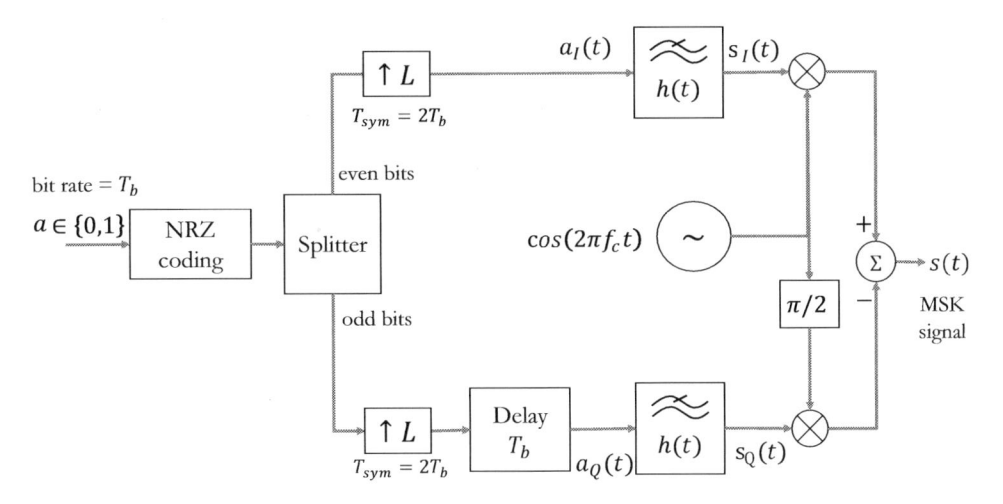

Fig. 2.24: A practical MSK modulator constructed according to the equivalence with OQPSK modulation

MSK demodulator

The corresponding demodulator is given in Figure 2.25 and the Python function for MSK demodulation (msk_demod) is also given next. Note that at the demodulator, the signal in the inphase and quadrature arms are multiplied by the absolute value of the half-cycle functions: $|cos\{\pi t/(2T_b)\}|$ and $|sin\{\pi t/(2T_b)\}|$, respectively. The rest of the demodulator is same as the OQPSK demodulator structure shown in Figure 2.15.

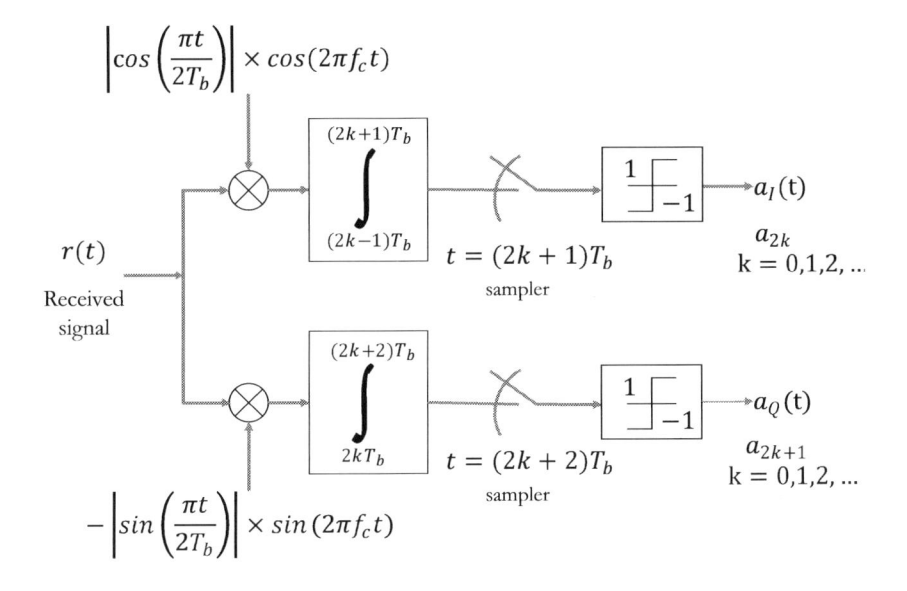

Fig. 2.25: MSK demodulator using the complex representation approach

Program 35: DigiCommPy\passband_modulations.py: MSK demodulator

```python
def msk_demod(r,N,fc,OF):
    """
    MSK demodulator
    Parameters:
        r : received signal at the receiver front end
        N : number of symbols transmitted
        fc : carrier frequency in Hertz
        OF : oversampling factor (at least 4 is better)
    Returns:
        a_hat : detected binary stream
    """
    L = 2*OF # samples in 2Tb duration
    Fs=OF*fc;Ts=1/Fs;Tb = OF*Ts; # sampling frequency, durations
    t=np.arange(-OF, len(r) - OF)/Fs # time base
    
    # cosine and sine functions for half-sinusoid shaping
    x=abs(np.cos(np.pi*t/(2*Tb)));y=abs(np.sin(np.pi*t/(2*Tb)))
    
    u=r*x*np.cos(2*np.pi*fc*t) # multiply I by half cosines and cos(2pifct)
    v=-r*y*np.sin(2*np.pi*fc*t) # multiply Q by half sines and sin(2pifct)
    
    iHat = np.convolve(u,np.ones(L)) # integrate for L (Tsym=2*Tb) duration
    qHat = np.convolve(v,np.ones(L)) # integrate for L (Tsym=2*Tb) duration
    
    iHat= iHat[L-1:-1-L:L]  # I- sample at the end of every symbol
    qHat= qHat[L+L//2-1:-1-L//2:L] # Q-sample from L+L/2th sample
    
    a_hat = np.zeros(N)
    a_hat[0::2] = iHat > 0 # thresholding - odd bits
    a_hat[1::2] = qHat > 0 # thresholding - even bits
    
    return a_hat
```

Performance simulation

The python script for simulating a complete communication chain with the aforementioned MSK modulator, AWGN channel and the MSK demodulator is given next (*msk.py*). The simulated timing waveforms at the transmitter are shown in Figure 2.26. The code also includes performance simulation of the MSK communication system over a range of E_b/N_0 values pertaining to the AWGN channel.

MSK modulator can be interpreted as having two BPSK-like modulators on the I-channel and Q-channel arms and the only difference being the half-symbol shift with the half-cycle sinusoidal shaping for MSK. Similar argument applies for MSK demodulator as well. Hence the theoretical bit error probability of MSK is identical to that of conventional BPSK (see Figure 2.6 for reference), which is given by

$$P_b = \frac{1}{2}\,erfc\left(\sqrt{\frac{E_b}{N_0}}\right) \tag{2.29}$$

2.8 Continuous Phase Modulation (CPM)

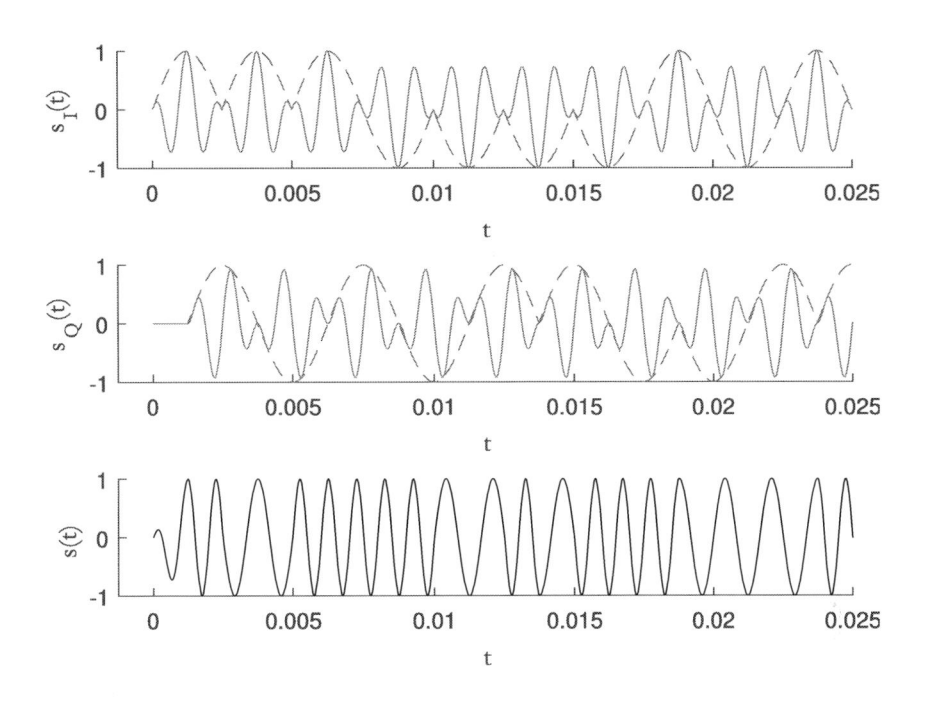

Fig. 2.26: Simulated MSK waveforms at the transmitter side

Program 36: DigiCommPy\chapter_2\msk.py: Performance of MSK over AWGN

```
#Execute in Python3: exec(open("chapter_2/msk.py").read())
import numpy as np #for numerical computing
import matplotlib.pyplot as plt #for plotting functions
from DigiCommPy.passband_modulations import msk_mod,msk_demod
from DigiCommPy.channels import awgn
from scipy.special import erfc

N = 100000 # Number of symbols to transmit
EbN0dB = np.arange(start=-4,stop = 11,step = 2) # Eb/N0 range in dB for simulation
fc = 800 # carrier frequency in Hertz
OF = 32 # oversampling factor, sampling frequency will be fs=OF*fc

BER = np.zeros(len(EbN0dB)) # For BER values for each Eb/N0

a = np.random.randint(2, size=N) # uniform random symbols from 0's and 1's
result = msk_mod(a,fc,OF,enable_plot=True) # MSK modulation
s = result['s(t)']

for i,EbN0 in enumerate(EbN0dB):
    # Compute and add AWGN noise
    r = awgn(s,EbN0,OF) # refer Chapter section 4.1
```

```
    a_hat = msk_demod(r,N,fc,OF) #receiver

    BER[i] = np.sum(a!=a_hat)/N # Bit Error Rate Computation

theoreticalBER = 0.5*erfc(np.sqrt(10**(EbN0dB/10))) # Theoretical bit error rate

#-------------Plots--------------------------
fig, ax = plt.subplots(nrows=1,ncols = 1)
ax.semilogy(EbN0dB,BER,'k*',label='Simulated') # simulated BER
ax.semilogy(EbN0dB,theoreticalBER,'r-',label='Theoretical')
ax.set_xlabel(r'$E_b/N_0$ (dB)')
ax.set_ylabel(r'Probability of Bit Error - $P_b$')
ax.set_title(['Probability of Bit Error for MSK modulation'])
ax.legend();fig.show();
```

2.9 Investigating phase transition properties

The phase transition properties of the different variants of QPSK schemes and MSK, are easily investigated using constellation plots. The next code sample demonstrates how to plot the signal space constellations, for the various modulations used in the transmitter.

Typically, in practical applications, the baseband modulated waveforms are passed through a pulse shaping filter for combating the phenomenon of intersymbol interference (ISI). The goal is to plot the constellation plots of various pulse-shaped baseband waveforms of the QPSK, O-QPSK and $\pi/4$-DQPSK schemes. A variety of pulse shaping filters are available and raised cosine filter is specifically chosen for this demo.

The *raised cosine* (RC) pulse comes with an adjustable transition band roll-off parameter α, using which the decay of the transition band can be controlled. The RC pulse shaping function is expressed in frequency domain as

$$P(f) = \begin{cases} T_{sym}, & |f| \leq \frac{1-\alpha}{2T_{sym}} \\ \frac{T_{sym}}{2}\left[1+\cos\left(\frac{\pi T_{sym}}{\alpha}\left[|f|-\frac{1-\alpha}{2T_{sym}}\right]\right)\right], & \frac{1-\alpha}{2T_{sym}} \leq f \leq \frac{1+\alpha}{2T_{sym}} \\ 0, & |f| \geq \frac{1+\alpha}{2T_{sym}} \end{cases} \quad (2.30)$$

Equivalently, in time domain, the impulse response corresponds to

$$p(t) = \frac{\sin\left(\frac{\pi t}{T_{sym}}\right)}{\frac{\pi t}{T_{sym}}} \frac{\cos\left(\frac{\pi \alpha t}{T_{sym}}\right)}{1-\left(\frac{2\alpha t}{T_{sym}}\right)^2} \quad \text{where } 0 \leq \alpha \leq 1 \quad (2.31)$$

A simple evaluation of the equation 2.31 produces singularities (undefined points) at $p(t = 0)$ and $p(t = \pm T_{sym}/(2\alpha))$. The value of the raised cosine pulse at these singularities can be obtained by applying L'Hospital's rule [8] and the values are

$$p(t=0) = 1 \quad (2.32)$$

$$p\left(t = \pm\frac{T_{sym}}{2\alpha}\right) = \frac{\alpha}{2}\sin\left(\frac{\pi}{2\alpha}\right) \quad (2.33)$$

Using the equations above, the raised cosine filter is implemented as a function in Python:

2.9 Investigating phase transition properties

Program 37: DigiCommPy\pulseshapers.py: Implementing a raised cosine pulse shaper

```python
def raisedCosineDesign(alpha, span, L):
    """
    Raised cosine FIR filter design
    Parameters:
        alpha : roll-off factor
        span : filter span in symbols
        L : oversampling factor (i.e, each symbol contains L samples)
    Returns:
        p - filter coefficients b of the designed
            FIR raised cosine filter
    """
    t = np.arange(-span/2, span/2 + 1/L, 1/L) # +/- discrete-time base
    with np.errstate(divide='ignore', invalid='ignore'):
        A = np.divide(np.sin(np.pi*t),(np.pi*t)) #assume Tsym=1
        B = np.divide(np.cos(np.pi*alpha*t),1-(2*alpha*t)**2)
        p = A*B
    #Handle singularities
    p[np.argwhere(np.isnan(p))] = 1 # singularity at p(t=0)
    # singularity at t = +/- Tsym/2alpha
    p[np.argwhere(np.isinf(p))] = (alpha/2)*np.sin(np.divide(np.pi,(2*alpha)))
    return p
```

The function is tested with the following code. It generates a raised cosine pulse for the given symbol duration $T_{sym} = 1s$ and plots the time-domain view and the frequency response as shown in Figure 2.27. From the plot, it can be observed that the RC pulse falls off at the rate of $1/|t|^3$ as $t \to \infty$, which is a significant improvement when compared to the decay rate of a sinc pulse which is $1/|t|$. It satisfies Nyquist criterion for zero ISI - the pulse hits zero crossings at desired sampling instants. The transition bands in the frequency domain can be made gradual (by controlling α) when compared to that of a sinc pulse.

Program 38: DigiCommPy\pulseshapers.py: Raised cosine pulses in time and frequency domain

```python
def raisedCosineDemo():
    """
    Raised Cosine pulses and their manifestation in frequency domain
    Usage:
        >> from DigiCommPy.pulseshapers import raisedCosineDemo
        >> raisedCosineDemo()
    """
    import matplotlib.pyplot as plt
    from scipy.fftpack import fft, fftshift
    
    Tsym = 1 # Symbol duration in seconds
    L = 32 # oversampling rate, each symbol contains L samples
    span = 10 # filter span in symbols
    alphas= [0, 0.3, 0.5, 1] # RC roll-off factors - valid range 0 to 1
    Fs = L/Tsym # sampling frequency
    
    lineColors = ['b','r','g','k']
    fig, (ax1,ax2) = plt.subplots(1, 2)
```

```python
    for i, alpha in enumerate(alphas):
        b = raisedCosineDesign(alpha,span,L) # RC Pulse design

        # time base for symbol duration
        t = Tsym* np.arange(-span/2, span/2 + 1/L, 1/L)

        # plot time domain view
        ax1.plot(t,b,lineColors[i],label=r'$\alpha$='+str(alpha))

        #Compute FFT and plot double sided frequency domain view
        NFFT = 1<<(len(b)-1).bit_length() #Set FFT length = nextpower2(len(b))
        vals = fftshift(fft(b,NFFT))
        freqs = Fs* np.arange(-NFFT/2,NFFT/2)/NFFT
        ax2.plot(freqs,abs(vals)/abs(vals[len(vals)//2]),\
                 lineColors[i],label=r'$\alpha$='+str(alpha))

    ax1.set_title('Raised cosine pulse');
    ax2.set_title('Frequency response')
    ax1.legend();ax2.legend()
    fig.show()
```

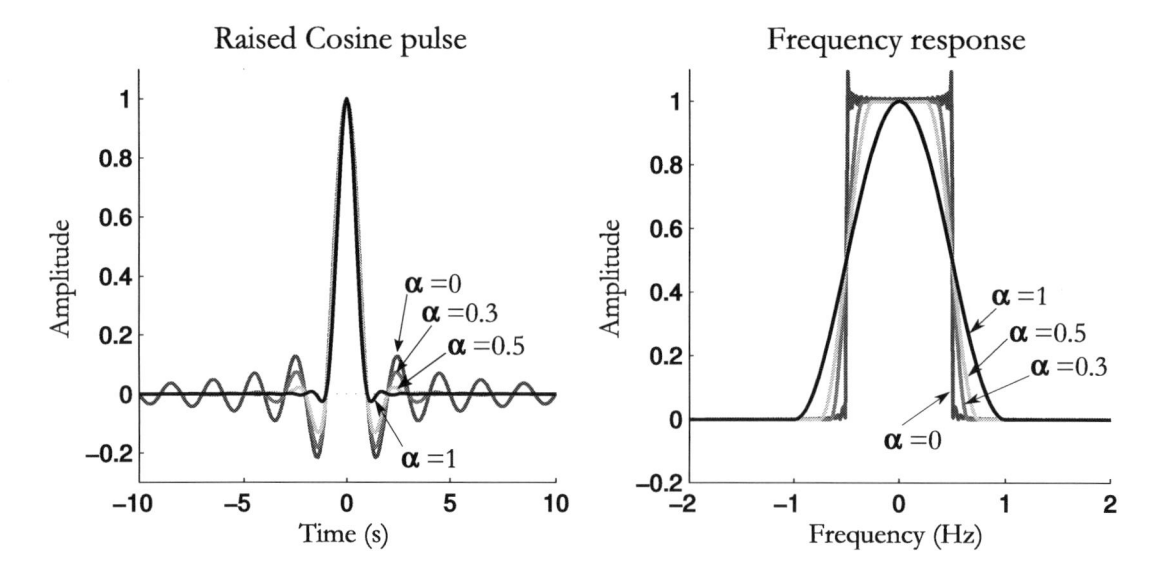

Fig. 2.27: Raised cosine pulse and its manifestation in frequency domain

Now that we have constructed a function for raised cosine pulse shaping filter, the next step is to generate modulated waveforms (using QPSK, O-QPSK and $\pi/4$-DQPSK schemes), pass them through a raised cosine filter having a roll-off factor, say $\alpha = 0.3$ and finally plot the constellation. The constellation for MSK modulated waveform is also plotted.

2.9 Investigating phase transition properties

Program 39: DigiCommPy\chapter_2\constellations.py: Constellations of RC filtered QPSK & MSK

```
#Execute in Python3: exec(open("chapter_2/constellations.py").read())
import numpy as np
import matplotlib.pyplot as plt
from DigiCommPy.passband_modulations import
    qpsk_mod,oqpsk_mod,piBy4_dqpsk_mod,msk_mod
from DigiCommPy.pulseshapers import raisedCosineDesign

N=1000 # Number of symbols to transmit, keep it small and adequate
fc=10; L=8 # carrier frequency and oversampling factor
a = np.random.randint(2, size=N) # uniform random symbols from 0's and 1's

#modulate the source symbols using QPSK,QPSK,pi/4-DQPSK and MSK
qpsk_result= qpsk_mod(a,fc,L)
oqpsk_result = oqpsk_mod(a,fc,L)
piby4qpsk_result = piBy4_dqpsk_mod(a,fc,L)
msk_result = msk_mod(a,fc,L);

#Pulse shape the modulated waveforms by convolving with RC filter
alpha = 0.3; span = 10 # RC filter alpha and filter span in symbols
b =raisedCosineDesign(alpha,span, L) # RC pulse shaper
iRC_qpsk=np.convolve(qpsk_result['I(t)'],b,mode='valid')#RC - QPSK I(t)
qRC_qpsk= np.convolve(qpsk_result['Q(t)'],b,mode='valid') #RC - QPSK Q(t)
iRC_oqpsk= np.convolve(oqpsk_result['I(t)'],b,mode='valid')#RC - OQPSK I(t)
qRC_oqpsk= np.convolve(oqpsk_result['Q(t)'],b,mode='valid')#RC - OQPSK Q(t)
iRC_piby4qpsk=np.convolve(piby4qpsk_result['U(t)'],b,mode='valid')#pi/4-QPSK I
qRC_piby4qpsk=np.convolve(piby4qpsk_result['V(t)'],b,mode='valid')#pi/4-QPSK Q
i_msk = msk_result['sI(t)'] # MSK sI(t)
q_msk = msk_result['sQ(t)'] # MSK sQ(t)

fig, axs = plt.subplots(2, 2)

axs[0,0].plot(iRC_qpsk,qRC_qpsk)# RC shaped QPSK
axs[0,1].plot(iRC_oqpsk,qRC_oqpsk)# RC shaped OQPSK
axs[1,0].plot(iRC_piby4qpsk,qRC_piby4qpsk)# RC shaped pi/4-QPSK
axs[1,1].plot(i_msk[20:-20],q_msk[20:-20])# RC shaped OQPSK

axs[0,0].set_title(r'QPSK, RC $\alpha$='+str(alpha))
axs[0,0].set_xlabel('I(t)');axs[0,0].set_ylabel('Q(t)');
axs[0,1].set_title(r'OQPSK, RC $\alpha$='+str(alpha))
axs[0,1].set_xlabel('I(t)');axs[0,1].set_ylabel('Q(t)');
axs[1,0].set_title(r'$\pi$/4 - QPSK, RC $\alpha$='+str(alpha))
axs[1,0].set_xlabel('I(t)');axs[1,0].set_ylabel('Q(t)');
axs[1,1].set_title('MSK')
axs[1,1].set_xlabel('I(t)');axs[1,1].set_ylabel('Q(t)');
fig.show()
```

The resulting simulated plot is shown in the Figure 2.28. From the resulting constellation plot, following conclusions can be reached.

- Conventional QPSK has 180° phase transitions and hence it requires linear amplifiers with high Q factor

- The phase transitions of Offset-QPSK are limited to 90° (the 180° phase transitions are eliminated)
- The signaling points for $\pi/4$-DQPSK is toggled between two sets of QPSK constellations that are shifted by 45° with respect to each other. Both the 90° and 180° phase transitions are absent in this constellation. Therefore, this scheme produces the lower envelope variations than the rest of the two QPSK schemes.
- MSK is a continuous phase modulation, therefore no abrupt phase transition occurs when a symbol changes. This is indicated by the smooth circle in the constellation plot. Hence, a bandlimited MSK signal will not suffer any envelope variation, whereas, the rest of the QPSK schemes suffer varied levels of envelope variations, when they are bandlimited.

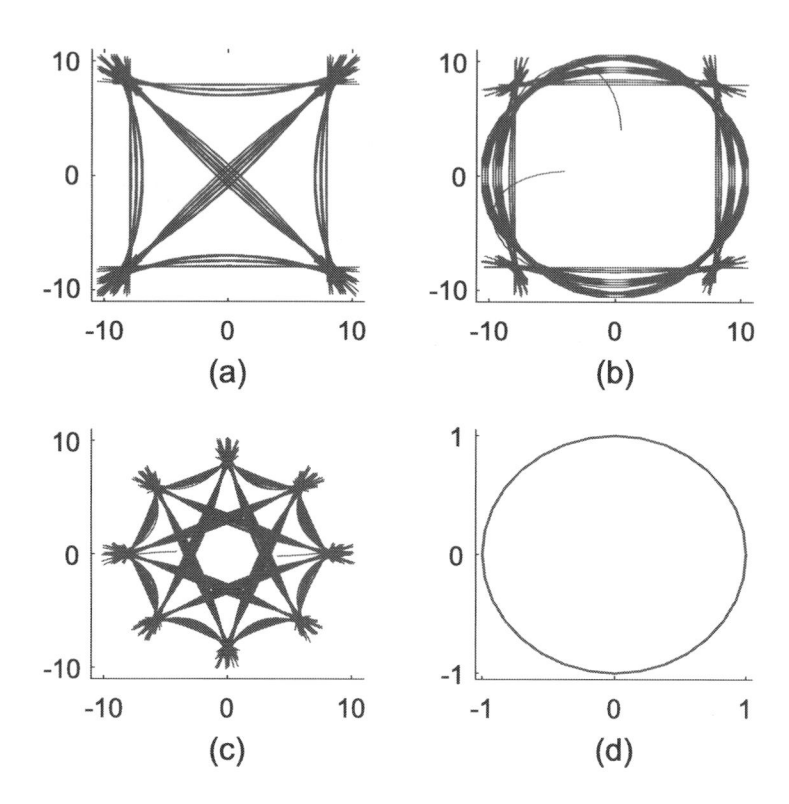

Fig. 2.28: Constellations plots for: (a) $\alpha = 0.3$ RC-filtered QPSK, (b) $\alpha = 0.3$ RC-filtered O-QPSK, (c) $\alpha = 0.3$ RC-filtered $\pi/4$-DQPSK and (d) MSK

2.10 Power spectral density (PSD) plots

Power spectral density (PSD) is a measure of a signal's power intensity in the frequency domain. Comparison of PSDs of various modulation schemes plays a vital role in understanding their spectral characteristics and to make decisions on choosing the modulation scheme that satisfies the requirements of a given communication system. A discussion on PSD estimates is given in chapter 1 section 1.4.

The function `plotWelchPSD`, given in chapter chapter 1 section 1.4, is utilized here for plotting the PSDs of BPSK, QPSK and MSK modulated signals with carrier frequency of $800Hz$ and the resulting plot is given in Figure 2.29.

2.10 Power spectral density (PSD) plots

Program 40: DigiCommPy\chapter_2\psd_estimates.py: PSD estimates of BPSK QPSK and MSK

```python
def bpsk_qpsk_msk_psd():
    # Usage:
    #    >> from chapter_2.psd_estimates import bpsk_qpsk_msk_psd
    #    >> bpsk_qpsk_msk_psd()
    from DigiCommPy.passband_modulations import bpsk_mod,qpsk_mod,msk_mod
    from DigiCommPy.essentials import plotWelchPSD
    N=100000 # Number of symbols to transmit
    fc=800;OF =8 # carrier frequency and oversamping factor
    fs = fc*OF # sampling frequency

    a = np.random.randint(2, size=N) # uniform random symbols from 0's and 1's
    (s_bb,t) = bpsk_mod(a,OF) # BPSK modulation(waveform) - baseband
    s_bpsk = s_bb*np.cos(2*np.pi*fc*t/fs) # BPSK with carrier
    s_qpsk = qpsk_mod(a,fc,OF)['s(t)'] # conventional QPSK
    s_msk = msk_mod(a,fc,OF)['s(t)'] # MSK signal

    # Compute and plot PSDs for each of the modulated versions
    fig, ax = plt.subplots(1, 1)
    plotWelchPSD(s_bpsk,fs,fc,ax = ax,color = 'b',label='BPSK')
    plotWelchPSD(s_qpsk,fs,fc,ax = ax,color = 'r',label='QPSK')
    plotWelchPSD(s_msk,fs,fc, ax = ax,color = 'k',label='MSK')
    ax.set_xlabel('$f-f_c$');ax.set_ylabel('PSD (dB/Hz)');ax.legend();fig.show()
```

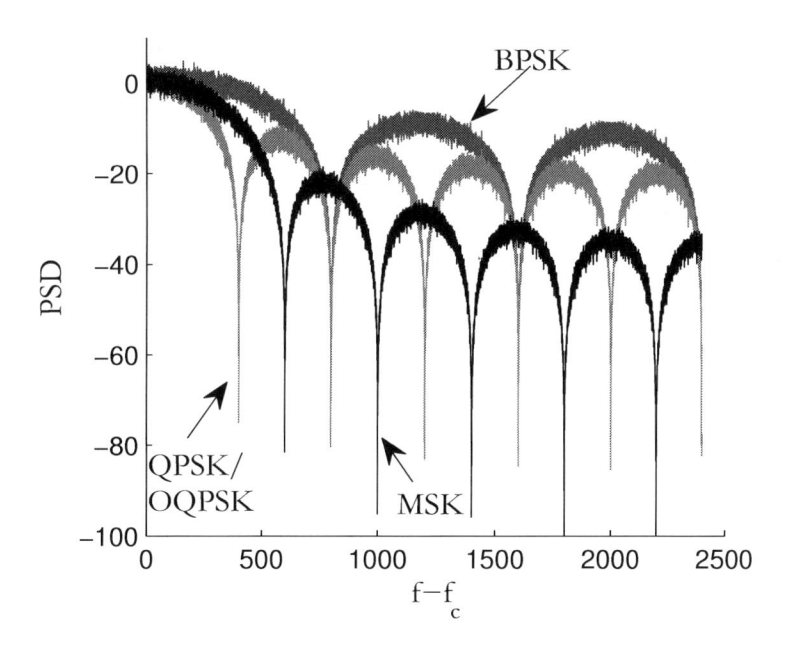

Fig. 2.29: PSD estimates for BPSK, QPSK and MSK signals

2.11 Gaussian Minimum Shift Keying (GMSK)

Minimum shift keying (MSK) is a special case of binary CPFSK with modulation index $h = 0.5$. It has features such as constant envelope, compact spectrum and good error rate performance. The fundamental problem with MSK is that the spectrum is not compact enough to satisfy the stringent requirements with respect to out-of-band radiation for technologies like GSM and DECT standard. These technologies have very high data rates approaching the RF channel bandwidth. A plot of MSK spectrum (Figure 2.29) will reveal that the sidelobes with significant energy, extend well beyond the transmission data rate. This is problematic, since it causes severe out-of-band interference in systems with closely spaced adjacent channels.

To satisfy such requirements, the MSK spectrum can be easily manipulated by using a pre-modulation low pass filter (LPF). The pre-modulation LPF should have the following properties and it is found that a Gaussian LPF will satisfy all of them [9].

- Sharp cut-off and narrow bandwidth - needed to suppress high frequency components.
- Lower overshoot in the impulse response - providing protection against excessive instantaneous frequency deviations.
- Preservation of filter output pulse area - thereby coherent detection can be applicable.

2.11.1 Pre-modulation Gaussian low pass filter

Gaussian Minimum Shift Keying (GMSK) is a modified MSK modulation technique, where the spectrum of MSK is manipulated by passing the rectangular shaped information pulses through a Gaussian LPF *prior* to the frequency modulation of the carrier. A typical Gaussian LPF, used in GMSK modulation standards, is defined by the zero-mean Gaussian (bell-shaped) impulse response.

$$h(t) = \frac{1}{\sqrt{2\pi\sigma^2}} exp\left(\frac{-t^2}{2\sigma^2}\right) \quad , \sigma^2 = \frac{ln(2)}{(2\pi B)^2} \qquad (2.34)$$

The parameter B is the 3-dB bandwidth of the LPF, which is determined from a parameter called BT_b as discussed next. If the input to the filter is an isolated unit rectangular pulse ($p(t) = 1, 0 \leq t \leq T_b$), the response of the filter will be [10]

$$g(t) = \frac{1}{2T_b}\left[Q\left(\frac{2\pi BT_b}{\sqrt{ln\,2}}\left(\frac{t}{T_b} - 1\right)\right) - Q\left(\frac{2\pi BT_b}{\sqrt{ln\,2}}\frac{t}{T_b}\right)\right] \qquad (2.35)$$

where,

$$Q(x) = \int_x^\infty \frac{1}{\sqrt{2\pi}} exp\left(-\frac{y^2}{2}\right) dy, \quad -\infty \leq t \leq \infty$$

It is important to note the distinction between the two equations - 2.34 and 2.35. The equation for $h(t)$ defines the impulse response of the LPF, whereas the equation for $g(t)$, also called as *frequency pulse shaping function*, defines the LPF's output when the filter gets excited with a rectangular pulse. This distinction is captured in Figure 2.30.

The aim of using GMSK modulation is to have a controlled MSK spectrum. Effectively, a variable parameter called BT_b, the product of 3-dB bandwidth of the LPF and the desired data-rate T_b, is often used by the designers to control the amount of spectrum efficiency required for the desired application. As a consequence, the 3-dB bandwidth of the aforementioned LPF is controlled by the BT_b design parameter. The range for the

2.11 Gaussian Minimum Shift Keying (GMSK)

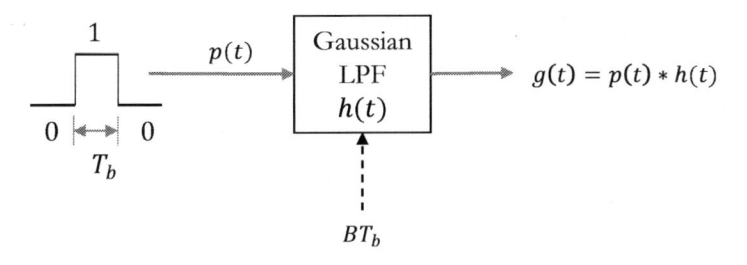

Fig. 2.30: Gaussian LPF: Relating $h(t)$ and $g(t)$

parameter BT_b is given as $0 < BT_b \leq \infty$. When $BT_b = \infty$, the impulse response $h(t)$ becomes a Dirac delta function $\delta(t)$, resulting in a transparent LPF and hence this configuration corresponds to MSK modulation.

The function to implement the Gaussian LPF's impulse response (equation 2.34), is given next. The Gaussian impulse response is of infinite duration and hence in digital implementations it has to be defined for a finite interval, as dictated by the function argument k in the code shown next. For example, in GSM standard, BT_b is chosen as 0.3 and the time truncation is done to three bit-intervals $k = 3$.

It is also necessary to normalize the filter coefficients of the computed LPF as

$$h[n] = \frac{h[n]}{\sum_{i=0}^{N-1} h[i]}, \quad n = 0, 1, ..., N-1 \tag{2.36}$$

Program 41: DigiCommPy\passband_modulations.py: Generate filter coefficients of Gaussian LPF

```python
def gaussianLPF(BT, Tb, L, k):
    """
    Generate filter coefficients of Gaussian low pass filter (used in gmsk_mod)
    Parameters:
        BT : BT product - Bandwidth x bit period
        Tb : bit period
        L : oversampling factor (number of samples per bit)
        k : span length of the pulse (bit interval)
    Returns:
        h_norm : normalized filter coefficients of Gaussian LPF
    """
    B = BT/Tb # bandwidth of the filter
    # truncated time limits for the filter
    t = np.arange(start = -k*Tb, stop = k*Tb + Tb/L, step = Tb/L)
    h = B*np.sqrt(2*np.pi/(np.log(2)))*np.exp(-2 * (t*np.pi*B)**2 /(np.log(2)))
    h_norm=h/np.sum(h)
    return h_norm
```

Based on the gaussianLPF function, given above, we can compute and plot the impulse response $h(t)$ and the response to an isolated unit rectangular pulse - $g(t)$. The resulting plot is shown in Figure 2.31.

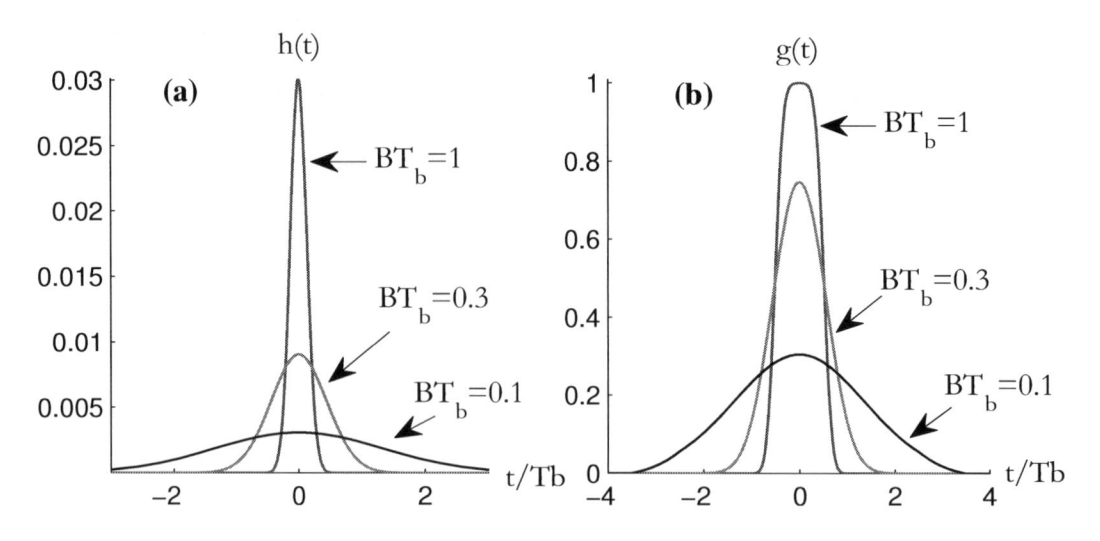

Fig. 2.31: Gaussian LPF:(a) impulse response and (b) response to isolated rectangular pulse

2.11.2 Quadrature implementation of GMSK modulator

Different implementations of a GMSK transmitter are possible. For conventional designs based on CPM representation, refer [10]. Quadrature design is another implementation for GMSK modulator that can be easily realized in software. The quadrature modulator for GMSK, shown in Figure 2.32, is readily obtained from Continuous Phase Modulation representation as

$$\begin{aligned}
s(t) &= cos\left(2\pi f_c t + \frac{\pi h}{T_b}\int_{-\infty}^{t} b(\tau)d\tau\right) \\
&= cos(2\pi f_c t) cos\left[\frac{\pi h}{T_b}\int_{-\infty}^{t} b(\tau)d\tau\right] - sin(2\pi f_c t) sin\left[\frac{\pi h}{T_b}\int_{-\infty}^{t} b(\tau)d\tau\right] \\
&= cos(2\pi f_c t) cos[\phi(t)] - sin(2\pi f_c t) sin[\phi(t)] \\
&= I(t)cos(2\pi f_c t) - Q(t)sin(2\pi f_c t)
\end{aligned} \qquad (2.37)$$

where, $b(t)$ is a Gaussian filtered NRZ data waveform sequence defined as

$$b(t) = \sum_{n=-\infty}^{\infty} a_n g(t - nT_b) \qquad (2.38)$$

It can be readily seen that the generated GMSK signal depends on two parameters:

- BT_b - the product of 3-dB LPF bandwidth and the bit-period
- h - the *modulation index* defined as the ratio of peak-to-peak frequency deviation and the bit rate

$$h = \frac{\text{peak-to-peak frequency deviation}}{bitrate} = 2\Delta f \times T_b \qquad (2.39)$$

GSM standard uses $BT_b = 0.3$ and $h = 0.5$ for generating the GMSK signal at the transmitter.

The function gmsk_mod implements the quadrature modulator (for $h = 0.5$). The simulated waveforms at various points in the modulator are given in Figure 2.33.

2.11 Gaussian Minimum Shift Keying (GMSK)

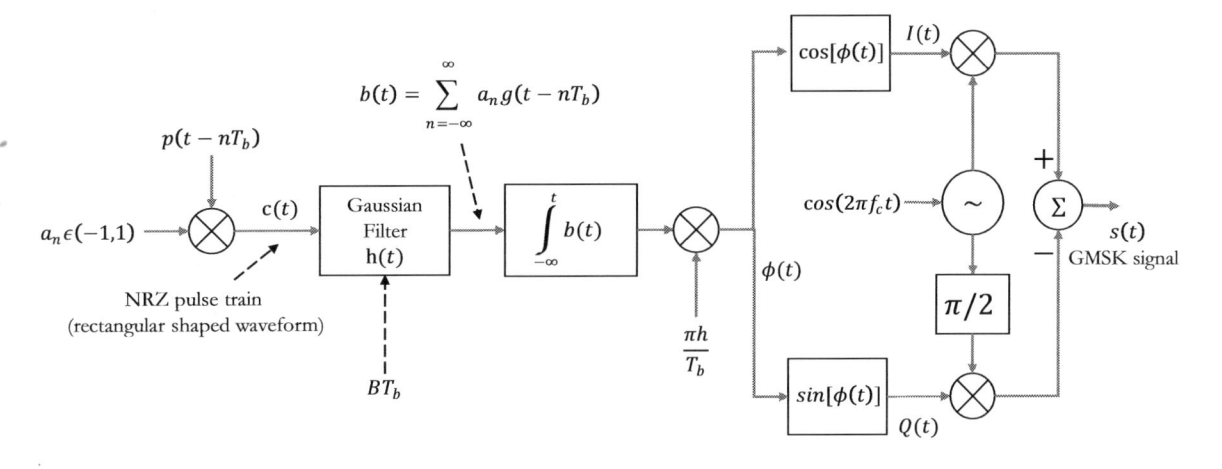

Fig. 2.32: Quadrature implementation of GMSK modulator

Program 42: DigiCommPy\passband_modulations.py: Implementation of GMSK modulator

```python
def gmsk_mod(a,fc,L,BT,enable_plot=False):
    """
    Function to modulate a binary stream using GMSK modulation
    Parameters:
        BT : BT product (bandwidth x bit period) for GMSK
        a : input binary data stream (0's and 1's) to modulate
        fc : RF carrier frequency in Hertz
        L : oversampling factor
        enable_plot: True = plot transmitter waveforms (default False)
    Returns:
        (s_t,s_complex) : tuple containing the following variables
            s_t : GMSK modulated signal with carrier s(t)
            s_complex : baseband GMSK signal (I+jQ)
    """
    from scipy.signal import upfirdn,lfilter
    
    fs = L*fc; Ts=1/fs;Tb = L*Ts; # derived waveform timing parameters
    c_t = upfirdn(h=[1]*L, x=2*a-1, up = L) #NRZ pulse train c(t)
    
    k=1 # truncation length for Gaussian LPF
    h_t = gaussianLPF(BT,Tb,L,k) # Gaussian LPF with BT=0.25
    b_t = np.convolve(h_t,c_t,'full') # convolve c(t) with Gaussian LPF to get b(t)
    bnorm_t = b_t/max(abs(b_t)) # normalize the output of Gaussian LPF to +/-1
    
    h = 0.5;
    # integrate to get phase information
    phi_t = lfilter(b = [1], a = [1,-1], x = bnorm_t*Ts) * h*np.pi/Tb
    
    I = np.cos(phi_t)
    Q = np.sin(phi_t) # cross-correlated baseband I/Q signals
```

```python
s_complex = I - 1j*Q # complex baseband representation
t = Ts* np.arange(start = 0, stop = len(I)) # time base for RF carrier
sI_t = I*np.cos(2*np.pi*fc*t); sQ_t = Q*np.sin(2*np.pi*fc*t)
s_t = sI_t - sQ_t # s(t) - GMSK with RF carrier

if enable_plot:
    fig, axs = plt.subplots(2, 4)
    axs[0,0].plot(np.arange(0,len(c_t))*Ts,c_t);
    axs[0,0].set_title('c(t)');axs[0,0].set_xlim(0,40*Tb)
    axs[0,1].plot(np.arange(-k*Tb,k*Tb+Ts,Ts),h_t);
    axs[0,1].set_title('$h(t): BT_b$='+str(BT))
    axs[0,2].plot(t,I,'--');axs[0,2].plot(t,sI_t,'r');
    axs[0,2].set_title('$I(t)cos(2 \pi f_c t)$');axs[0,2].set_xlim(0,10*Tb)
    axs[0,3].plot(t,Q,'--');axs[0,3].plot(t,sQ_t,'r');
    axs[0,3].set_title('$Q(t)sin(2 \pi f_c t)$');axs[0,3].set_xlim(0,10*Tb)
    axs[1,0].plot( np.arange(0,len(bnorm_t))*Ts,bnorm_t);
    axs[1,0].set_title('b(t)');axs[1,0].set_xlim(0,40*Tb)
    axs[1,1].plot(np.arange(0,len(phi_t))*Ts, phi_t);
    axs[1,1].set_title('$\phi(t)$')
    axs[1,2].plot(t,s_t);axs[1,2].set_title('s(t)');
    axs[1,2].set_xlim(0,20*Tb)
    axs[1,3].plot(I,Q);axs[1,3].set_title('constellation')
    fig.show()
return (s_t,s_complex)
```

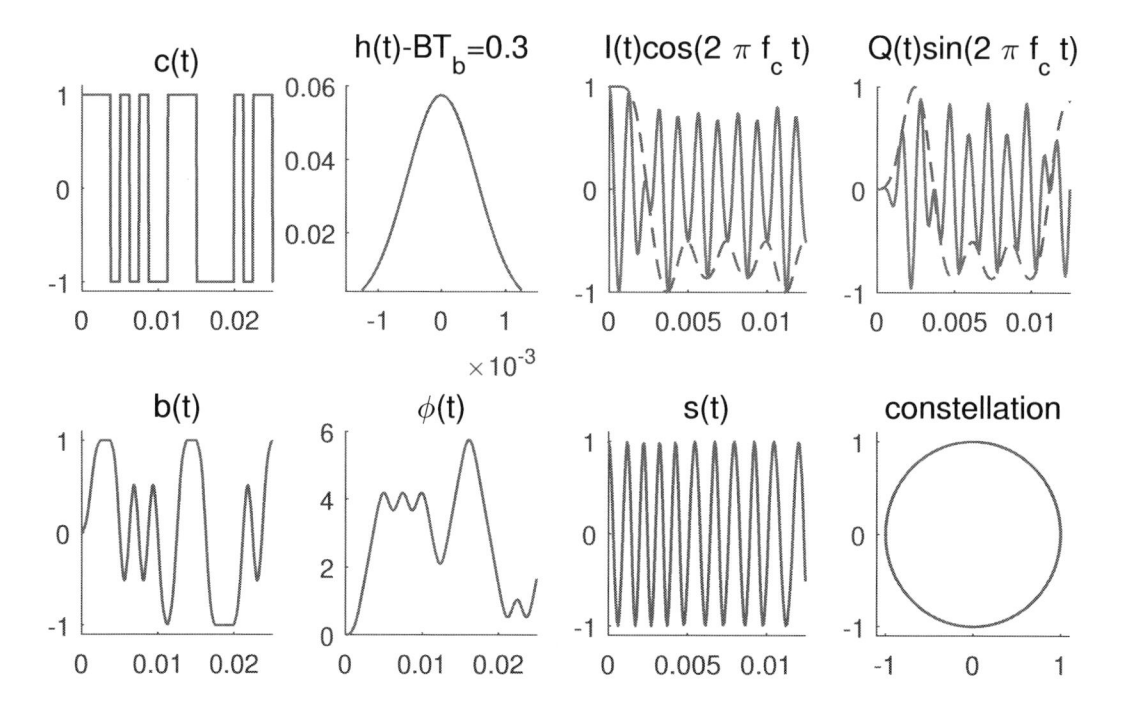

Fig. 2.33: Baseband waveforms at various points of the GMSK modulator

2.11 Gaussian Minimum Shift Keying (GMSK)

2.11.3 GMSK spectra

The PSD of GMSK signal is computed using the windowed Welch spectrum estimation method as described in chapter chapter 1 section 1.4. The PSD of the GMSK signal for various BT_b values of the Gaussian LPF, are shown in Figure 2.34. For relatively small values of BT_b products (≤ 0.3), the truncation of the LPF coefficients leads to spectrum floor. This result matches the documented observation in [11].

Program 43: DigiCommPy\chapter_2\psd_estimates.py: GMSK PSD - Welch spectrum estimation

```
def gmsk_psd():
    from DigiCommPy.passband_modulations import gmsk_mod
    from DigiCommPy.essentials import plotWelchPSD

    N = 10000 # Number of symbols to transmit
    fc = 800 # carrier frequency in Hertz
    L = 16 # oversampling factor,use L= Fs/Fc, where Fs >> 2xFc
    fs = L*fc
    a = np.random.randint(2, size=N) # uniform random symbols from 0's and 1's

    #'_':unused output variable
    (s1 , _ ) = gmsk_mod(a,fc,L,BT=0.3, enable_plot=True) # BT_b=0.3
    (s2 , _ ) = gmsk_mod(a,fc,L,BT=0.5) # BT_b=0.5
    (s3 , _ ) = gmsk_mod(a,fc,L,BT=0.7) # BT_b=0.7
    (s4 , _ ) = gmsk_mod(a,fc,L,BT=10000) # BT_b=very value value (MSK)

    # Compute and plot PSDs for each of the modulated versions
    fig, ax = plt.subplots(1, 1)
    plotWelchPSD(s1,fs,fc, ax = ax , color = 'r', label = '$BT_b=0.3$')
    plotWelchPSD(s2,fs,fc, ax = ax , color = 'b', label = '$BT_b=0.5$')
    plotWelchPSD(s3,fs,fc, ax = ax , color = 'm', label = '$BT_b=0.7$')
    plotWelchPSD(s4,fs,fc, ax = ax , color = 'k', label = '$BT_b=\infty$')
    ax.set_xlabel('$f-f_c$'); ax.set_ylabel('PSD (dB/Hz)')
    ax.legend(); fig.show()
```

2.11.4 GMSK demodulator

There exists numerous methods to demodulate a GMSK signal. A good survey of them can be found in the references [11] and [12]. A cross-coupled IQ coherent demodulator, shown in Figure 2.35, is studied here for the simulation. The received signal, $r(t)$, is first converted to an intermediate frequency (IF) using a band pass filter (not shown in the figure). With the help of recovered carrier, the IF signal is translated to baseband using the IQ implementation. The LPF filters in the inphase and quadrature arms, are simply used to remove the $2f_{if}$ frequency components that result from the down conversion multipliers. The baseband signals $I(t)$ and $Q(t)$ can be used by the decision algorithm (MLSE, parallel/serial type MSK detection etc..,) for data detection.

From the modulator structure shown in Figure 2.32, it can be readily seen that the baseband inphase and quadrature components are correlated as

$$\begin{aligned} I(t) &= cos[\phi(t)] \\ Q(t) &= sin[\phi(t)] \end{aligned} \tag{2.40}$$

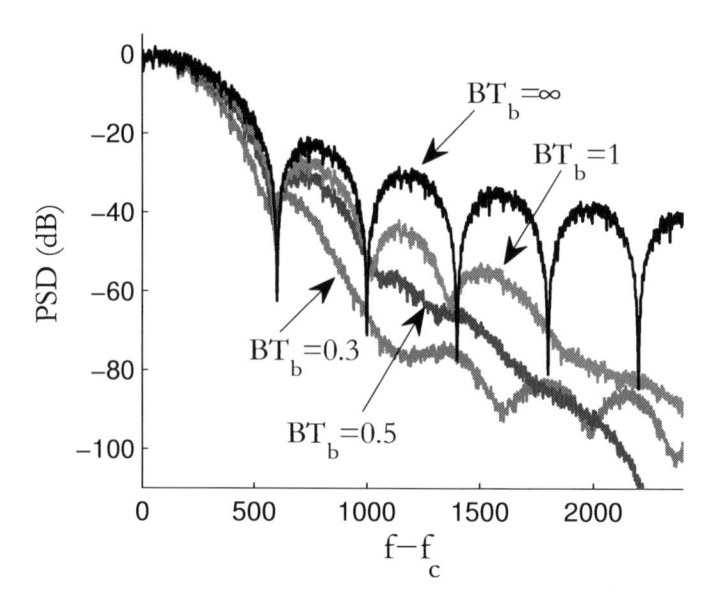

Fig. 2.34: GMSK spectrum using truncated impulse response (k=1 symbol)

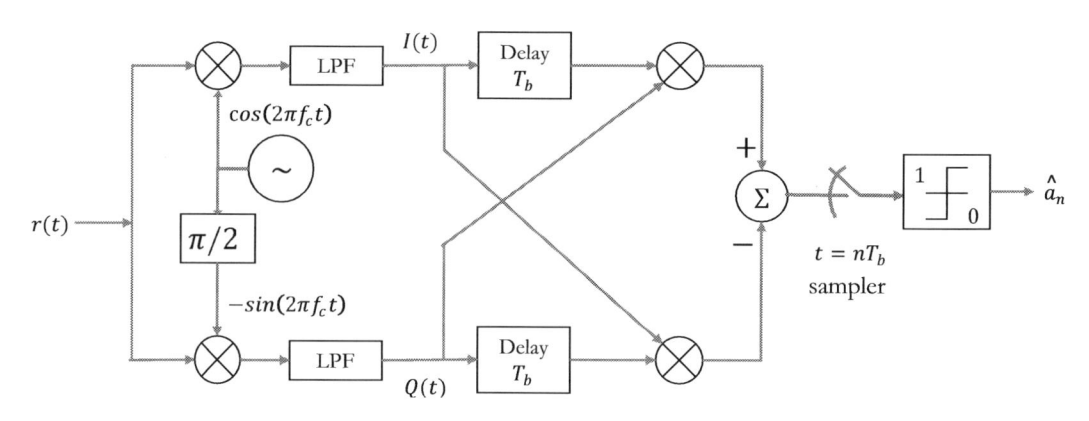

Fig. 2.35: Cross-coupled baseband IQ detector for GMSK demodulation

At this stage, it is tempting to deduce the phase information $\phi(t)$ directly from the baseband quadrature components as

$$\phi(t) = tan^{-1}\left(\frac{Q(t)}{I(t)}\right) = ARCTAN2\left(\frac{Q(t)}{I(t)}\right) \qquad (2.41)$$

Though, the above equation is mathematically correct, applying this direct approach for the detection at the receiver, is not a good solution, especially when the received signal is mixed with noise. Due to the addition of noise, any slight deviation in the detected instantaneous phase, leads to error propagation. Since the GMSK is a continuous phase modulation scheme, the correct method is to find the phase difference over a single bit interval and then build a decision logic using that phase difference.

The phase difference over one bit interval (T_b) is given as

$$\Delta\phi(t) = \phi(t) - \phi(t - T_b) \qquad (2.42)$$

2.11 Gaussian Minimum Shift Keying (GMSK)

Taking $sin()$ on both sides and applying trigonometric identity for $sin(a-b)$,

$$\begin{aligned}sin[\Delta\phi(t)] &= sin[\phi(t)-\phi(t-T_b)] \\ &= sin[\phi(t)]cos[\phi(t-T_b)] - cos[\phi(t)]sin[\phi(t-T_b)] \\ &= Q(t)I(t-T_b) - I(t)Q(t-T_b) \end{aligned} \quad (2.43)$$

The data bits are decided by applying hard decision on the sign of the above cross-coupled term (now you can relate how the demodulator in Figure 2.35 got the cross-coupled structure).

The following function gmsk_demod implements the cross-coupled detector using equation 2.43. It operates on the baseband received signal represented in complex IQ form.

Program 44: DigiCommPy\passband_modulations.py: Implementation of GMSK demodulator

```
def gmsk_demod(r_complex,L):
    """
    Function to demodulate a baseband GMSK signal
    Parameters:
        r_complex : received signal at receiver front end (complex form - I+jQ)
        L : oversampling factor
    Returns:
        a_hat : detected binary stream
    """
    I=np.real(r_complex); Q = -np.imag(r_complex); # I,Q streams
    z1 = Q * np.hstack((np.zeros(L), I[0:len(I)-L]))
    z2 = I * np.hstack((np.zeros(L), Q[0:len(I)-L]))
    z = z1 - z2
    a_hat = (z[2*L-1:-L:L] > 0).astype(int) # sampling and hard decision
    #sampling indices depend on the truncation length (k) of Gaussian LPF defined in
    ↪   the modulator
    return a_hat
```

2.11.5 Performance

The error rate performance of the baseband IQ modulator and cross-coupled demodulator over an AWGN channel is simulated and the results are plotted in Figure 2.36. Further improvements are possible when other techniques like MLSE (Viterbi algorithm), equalization etc are applied for detection [12]. Describing all the possible methods for demodulation is beyond the scope of this text.

Program 45: DigiCommPy\chapter_2\gmsk.py: Performance simulation of baseband GMSK

```
#Execute in Python3: exec(open("chapter_2/gmsk.py").read())
import numpy as np #for numerical computing
import matplotlib.pyplot as plt #for plotting functions
from DigiCommPy.passband_modulations import gmsk_mod,gmsk_demod
from DigiCommPy.channels import awgn

N=100000 # Number of symbols to transmit
EbN0dB = np.arange(start=0,stop = 19, step = 2) # Eb/N0 range in dB for simulation
BTs = [0.1, 0.3 ,0.5, 1] # Gaussian LPF's BT products
```

```
fc = 800 # Carrier frequency in Hz (must be < fs/2 and > fg)
L = 16 # oversampling factor

fig, axs = plt.subplots(nrows=1,ncols = 1)
lineColors = ['g','b','k','r']

for i,BT in enumerate(BTs):
    a = np.random.randint(2, size=N) # uniform random symbols from 0's and 1's
    (s_t,s_complex) = gmsk_mod(a,fc,L,BT) # GMSK modulation
    BER = np.zeros(len(EbN0dB)) # For BER values for each Eb/N0

    for j,EbN0 in enumerate(EbN0dB):
        r_complex = awgn(s_complex,EbN0) # refer Chapter section 4.1
        a_hat = gmsk_demod(r_complex,L) # Baseband GMSK demodulation
        BER[j] = np.sum(a!=a_hat)/N # Bit Error Rate Computation

    axs.semilogy(EbN0dB,BER,lineColors[i]+'*-',label='$BT_b=$'+str(BT))

axs.set_title('Probability of Bit Error for GMSK modulation')
axs.set_xlabel('E_b/N_0 (dB)');axs.set_ylabel('Probability of Bit Error - $P_b$')
axs.legend();fig.show()
```

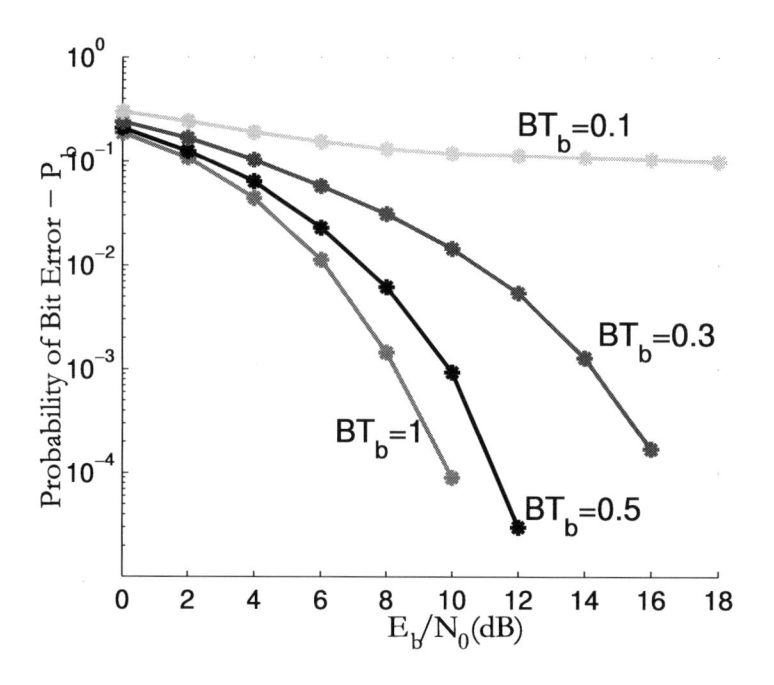

Fig. 2.36: BER performance of GMSK IQ mod-demod (with Gaussian LPF response truncated to k=1 symbol)

2.12 Frequency Shift Keying (FSK)

FSK is a frequency modulation scheme in which the digital signal is modulated as discrete variations in the frequency of the carrier signal. The simplest FSK configuration, called *Binary FSK* (BFSK), uses two discrete frequencies to represent the binary data, one frequency for representing binary 0 and another frequency for representing binary 1. When extended to more frequencies it is called *M-ary FSK* or simply MFSK. Waveform simulation of BFSK modulator-demodulator, is discussed in this section. For simulating MFSK modulation-demodulation refer section 3.4.5 in chapter 3. Performance simulation of MFSK transmission is discussed in chapter 4.

2.12.1 Binary-FSK (BFSK)

In binary-FSK, the binary data is transmitted using two waveforms of frequencies f_1 and f_2 - that are slightly offset by Δf from the center of the carrier frequency f_c.

$$f_1 = f_c + \frac{\Delta f}{2}$$
$$f_2 = f_c - \frac{\Delta f}{2} \qquad (2.44)$$

Sometimes, the frequency separation Δf is normalized to the bit-rate T_b using the *modulation index* as

$$h = \Delta f T_b \qquad (2.45)$$

The BFSK waveform is generated as

$$S_{BFSK}(t) = \begin{cases} A\cos[2\pi f_1 t + \phi_1] = A\cos\left[2\pi\left(f_c + \frac{\Delta f}{2}\right)t + \phi_1\right], & 0 \leq t \leq T_b, \; for\; binary\; 1 \\ A\cos[2\pi f_2 t + \phi_2] = A\cos\left[2\pi\left(f_c - \frac{\Delta f}{2}\right)t + \phi_2\right], & 0 \leq t \leq T_b, \; for\; binary\; 0 \end{cases} \qquad (2.46)$$

Here, ϕ_1 and ϕ_2 are the initial phases of the carrier at $t = 0$. If $\phi_1 = \phi_2$, then the modulation is called *coherent* BFSK, otherwise it is called *non-coherent* BFSK. From equations 2.45 and 2.46, the binary FSK scheme is represented as

$$S_{BFSK}(t) = \begin{cases} A\cos\left[\left(2\pi f_c + \frac{\pi h}{T_b}\right)t + \phi_1\right], & 0 \leq t \leq T_b, \; for\; binary\; 1 \\ A\cos\left[\left(2\pi f_c - \frac{\pi h}{T_b}\right)t + \phi_2\right], & 0 \leq t \leq T_b, \; for\; binary\; 0 \end{cases} \qquad (2.47)$$

The basis functions of the BFSK can be made orthogonal in signal space by the proper selection of the modulation index h or the frequency separation Δf.

2.12.2 Orthogonality condition for non-coherent BFSK detection

Often, the frequencies f_1 and f_2 are chosen to be orthogonal. The choice of non-coherent or coherent FSK generation affects the orthogonality between the chosen frequencies. The orthogonality condition for non-coherent FSK ($\phi_1 \neq \phi_2$) is determined first and the results can be used to find the condition for coherent FSK ($\phi_1 = \phi_2 = \phi$).

The condition for orthogonality can be obtained by finding the correlation between the two waveforms with respect to the frequency shift Δf between the two frequencies or with respect to the modulation index h. The two frequencies are said to be orthogonal at places where the cross-correlation function equates to zero.

$$R(f_1, f_2) = R(\Delta f) = \int_0^{T_b} cos(2\pi f_1 t + \phi_1) cos(2\pi f_2 t + \phi_2) dt = 0 \tag{2.48}$$

Applying trigonometric identities, the correlation function can be written as

$$\frac{1}{2} \int_0^{T_b} cos(2\pi[f_1 + f_2]t + [\phi_1 + \phi_2]) dt + \frac{1}{2} \int_0^{T_b} cos(2\pi[f_1 - f_2]t + [\phi_1 - \phi_2]) dt = 0$$

$$\int_0^{T_b} \left[cos(2\pi[f_1 + f_2]t) cos(\phi_1 + \phi_2) - sin(2\pi[f_1 + f_2]t) sin(\phi_1 + \phi_2) \right] dt$$

$$+ \int_0^{T_b} \left[cos(2\pi[f_1 - f_2]t) cos(\phi_1 - \phi_2) - sin(2\pi[f_1 - f_2]t) sin(\phi_1 - \phi_2) \right] dt = 0 \tag{2.49}$$

Further expanding the terms and applying the limits after integration

$$cos(\phi_1 + \phi_2) \frac{sin(2\pi[f_1 + f_2]T_b)}{2\pi[f_1 + f_2]} + sin(\phi_1 + \phi_2) \left[\frac{cos(2\pi[f_1 + f_2]T_b) - 1}{2\pi[f_1 + f_2]} \right]$$

$$+ cos(\phi_1 - \phi_2) \frac{sin(2\pi[f_1 - f_2]T_b)}{2\pi[f_1 - f_2]} + sin(\phi_1 - \phi_2) \left[\frac{cos(2\pi[f_1 - f_2]T_b) - 1}{2\pi[f_1 - f_2]} \right] = 0 \tag{2.50}$$

The above equation can be satisfied, if and only if, for some positive integers m and n,

$$2\pi[f_1 + f_2]T_b = 2m\pi \quad , m = 1, 2, 3, \ldots$$
$$2\pi[f_1 - f_2]T_b = 2n\pi \quad , n = 1, 2, 3, \ldots \tag{2.51}$$

This leads to the following solution.

$$f_1 = \frac{m+n}{2T_b}, \quad f_2 = \frac{m-n}{2T_b} \tag{2.52}$$

Therefore, the frequency separation for non-coherent case ($\phi_1 \neq \phi_2$) is given by

$$\Delta f = f_1 - f_2 = \frac{n}{T_b} \quad , n = 1, 2, 3, \ldots \tag{2.53}$$

Hence the minimum frequency separation for non-coherent FSK is obtained when $n = 1$. When expressed in terms of modulation index, using equation 2.45), the minimum frequency separation for obtaining orthogonal tones in the case of non-coherent FSK, occurs when $h = 1$.

$$(\Delta f)_{min} = \frac{1}{T_b} \Rightarrow h_{min} = 1 \tag{2.54}$$

2.12.3 Orthogonality condition for coherent BFSK

Noting that for coherent FSK generation, the initial phases are same for all frequency waveforms (i.e, $\phi_1 = \phi_2 = \phi$), the equation 2.50 shrinks to

$$cos(2\phi)\frac{sin(2\pi[f_1+f_2]T_b)}{2\pi[f_1+f_2]} + sin(2\phi)\left[\frac{cos(2\pi[f_1+f_2]T_b)-1}{2\pi[f_1+f_2]}\right] + \frac{sin(2\pi[f_1-f_2]T_b)}{2\pi[f_1-f_2]} = 0 \quad (2.55)$$

The above equation can be satisfied, if and only if, for some positive integers m and n,

$$2\pi[f_1+f_2]T_b = 2m\pi \quad, m = 1,2,3,...$$
$$2\pi[f_1-f_2]T_b = n\pi \quad, n = 1,2,3,... \quad (2.56)$$

This leads to

$$f_1 = \frac{2m+n}{4T_b}, \quad f_2 = \frac{2m-n}{4T_b}$$

Therefore, the frequency separation for coherent FSK ($\phi_1 = \phi_2$) is given by

$$\Delta f = f_1 - f_2 = \frac{n}{2T_b} \quad, n = 1,2,3,... \quad (2.57)$$

Hence the minimum frequency separation for coherently detected FSK is obtained when $n = 1$. When expressed in terms of modulation index, using equation 2.45, the minimum frequency separation required for obtaining orthogonal tones for the case of coherent FSK, occurs when $h = 0.5$, (which corresponds to *minimum shift keying* (MSK) should the phase change be continuous).

$$(\Delta f)_{min} = \frac{1}{2T_b} \Rightarrow h_{min} = 0.5 \quad (2.58)$$

Minimum shift keying (MSK) is a particular form of coherent FSK that not only guarantees minimum frequency separation but also provides continuous phase at the bit transitions. Refer section 2.8.3 on MSK for more details.

It is evident from equations 2.54 and 2.58, while maintaining the orthogonality between the carrier tones, given the same bit rate T_b, coherent FSK occupies less bandwidth when compared to the non-coherently detected FSK. Thus, coherently detected FSK is more *bandwidth efficient* when compared to non-coherently detected FSK.

Table 2.2, captures the difference between coherent and non-coherent BFSK. The same concepts can be extended to M-ary FSK (MFSK).

2.12.4 Modulator

The discrete time equivalent model for generating orthogonal BFSK signal is shown in Figure 2.37. Here, the discrete time sampling frequency (f_s) is chosen high enough when compared to the carrier frequency (f_c). The number of discrete samples in a bit-period L is chosen such that each bit-period is represented by sufficient number of samples to accommodate a few cycles of the frequencies of f_1 and f_2. The frequency separation Δf between f_1 and f_2 is computed from equation 2.45.

The function bfsk_mod implements the discrete-time equivalent model for BFSK generation, shown in Figure 2.37. It supports generating both coherent and non-coherent forms of BFSK signal. If the coherent

	Non-coherent BFSK	Coherent BFSK
Initial phases during waveform generation	$\phi_1 \neq \phi_2$	$\phi_1 = \phi_2$
Orthogonality condition 1	f_1, f_2 must be integer multiples of $1/(2T_b)$	f_1, f_2 must be integer multiples of $1/(4T_b)$
Orthogonality condition 2	The difference $\Delta f = f_1 - f_2$ must be an integer multiple of $1/T_b$	The difference $\Delta f = f_1 - f_2$ must be an integer multiple of $1/(2T_b)$
Minimum frequency separation	$1/T_b$	$1/(2T_b)$
Modulation index for minimum frequency separation	$h = 1$	$h = 0.5$ (MSK)
Demodulation	can only be non-coherently demodulated	demodulation could be coherent or non-coherent
Performance in AWGN when coherently demodulated	cannot apply coherent demodulation	$Q\left(\sqrt{E_b/N_0}\right)$
Performance in AWGN when non-coherently demodulated	$0.5 \, e^{-E_b/(2N_0)}$	$0.5 \, e^{-E_b/(2N_0)}$

Table 2.2: Summary of coherent and non-coherent BFSK schemes

BFSK is chosen, the function returns the generated initial random phase ϕ, in addition to the generated BFSK signal. This phase information is crucial for coherent detection at the receiver.

The coherent demodulator is described in the next section. The phase information generated by the coherent modulator should be passed to the coherent demodulator function (bfsk_coherent_demod) shown in 2.12.5. If non-coherent BFSK is chosen at the modulator, the phase argument returned by the function is irrelevant at the receiver. Moreover, a coherently modulated BFSK signal can also be non-coherently demodulated. The non-coherent demodulator is given in section 2.12.6.

Program 46: DigiCommPy\passband_modulations.py: Coherent & non-coherent discrete-time BFSK

```
def bfsk_mod(a,fc,fd,L,fs,fsk_type='coherent',enable_plot = False):
    """
    Function to modulate an incoming binary stream using BFSK
    Parameters:
        a : input binary data stream (0's and 1's) to modulate
        fc : center frequency of the carrier in Hertz
        fd : frequency separation measured from Fc
        L : number of samples in 1-bit period
        fs : Sampling frequency for discrete-time simulation
        fsk_type : 'coherent' (default) or 'noncoherent' FSK generation
        enable_plot: True = plot transmitter waveforms (default False)
    Returns:
        (s_t,phase) : tuple containing following parameters
            s_t : BFSK modulated signal
            phase : initial phase generated by modulator, applicable only for
    coherent FSK. It can be used when using coherent detection at Rx
    """
    from scipy.signal import upfirdn
    a_t = upfirdn(h=[1]*L, x=a, up = L) #data to waveform
```

2.12 Frequency Shift Keying (FSK)

```
    t = np.arange(start=0,stop=len(a_t))/fs #time base
    if fsk_type.lower() == 'noncoherent':
        # carrier 1 with random phase
        c1 = np.cos(2*np.pi*(fc+fd/2)*t+2*np.pi*np.random.random_sample())
        # carrier 2 with random phase
        c2 = np.cos(2*np.pi*(fc-fd/2)*t+2*np.pi*np.random.random_sample())
    else: #coherent is default
        # random phase from uniform distribution [0,2pi)
        phase=2*np.pi*np.random.random_sample()
        c1 = np.cos(2*np.pi*(fc+fd/2)*t+phase) # carrier 1 with random phase
        c2 = np.cos(2*np.pi*(fc-fd/2)*t+phase) # carrier 2 with the same random phase
    s_t = a_t*c1 +(-a_t+1)*c2 # BFSK signal (MUX selection)

    if enable_plot:
        fig, (ax1,ax2)=plt.subplots(2, 1);ax1.plot(t,a_t);ax2.plot(t,s_t);fig.show()
    return (s_t,phase)
```

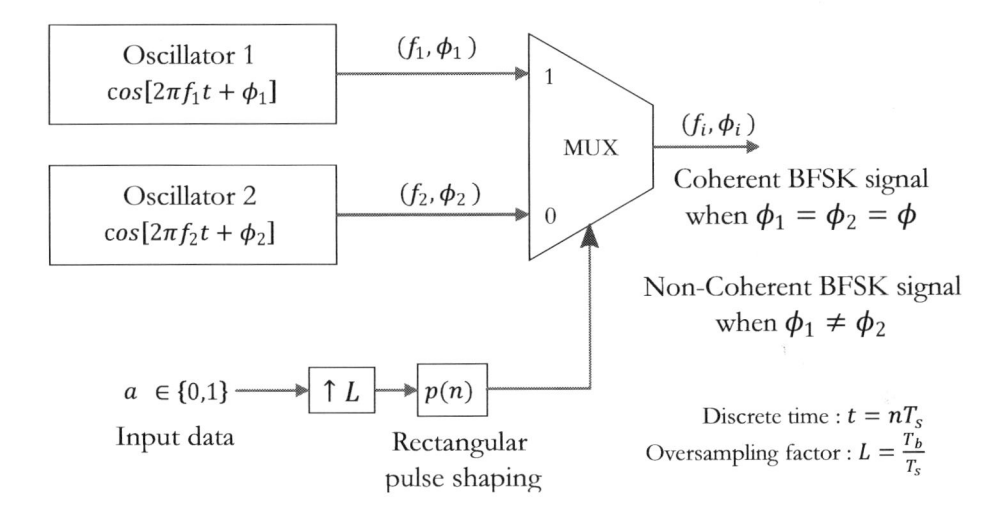

Fig. 2.37: Discrete-time equivalent model for BFSK modulation

2.12.5 Coherent demodulator

There are several ways to demodulate a coherent FSK signal and the *Voltage Controlled Oscillator* (VCO) based demodulator is the most popular among them. The coherent demodulator for FSK can also be implemented using only one correlator, and its discrete-time equivalent model is shown in Figure 2.38. Here, the demodulator requires the initial phase information ($\phi_1 = \phi_2 = \phi$) that was used to generate the coherent FSK signal at the transmitter. If this phase information is not accurately known at the receiver, the demodulator fails completely. In order to avoid carrier phase recovery, a non-coherent demodulator can be used to demodulate a coherent FSK signal, but there is an error rate performance penalty in doing so (see next section).

r(t) → ⊗ → Σ(·) → ↓L → > 0 ? → $\hat{a}$

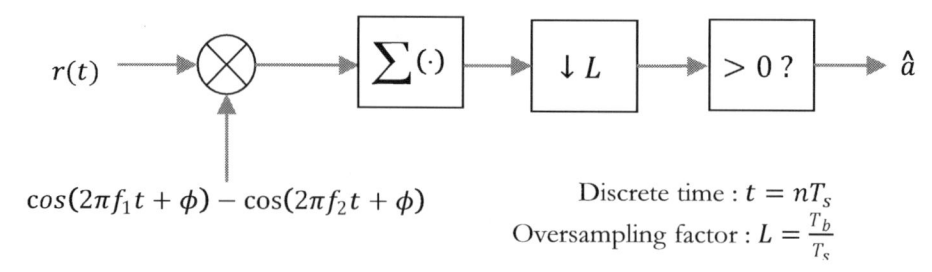

$\cos(2\pi f_1 t + \phi) - \cos(2\pi f_2 t + \phi)$

Discrete time : $t = nT_s$
Oversampling factor : $L = \dfrac{T_b}{T_s}$

Fig. 2.38: Discrete-time equivalent model for coherent demodulation of coherent FSK

The theoretical bit-error probability of a coherently detected coherent binary FSK, over AWGN noise, where two carrier signals are orthogonal and equi-probable, is given by

$$P_b = Q\left(\sqrt{\dfrac{E_b}{N_0}}\right) = \dfrac{1}{2} erfc\left(\sqrt{\dfrac{E_b}{2N_0}}\right) \tag{2.59}$$

Program 47: DigiCommPy\passband_modulations.py: Coherent demodulator for coherent BFSK

```python
def bfsk_coherent_demod(r_t,phase,fc,fd,L,fs):
    """
    Coherent demodulation of BFSK modulated signal
    Parameters:
        r_t : BFSK modulated signal at the receiver r(t)
        phase : initial phase generated at the transmitter
        fc : center frequency of the carrier in Hertz
        fd : frequency separation measured from Fc
        L : number of samples in 1-bit period
        fs : Sampling frequency for discrete-time simulation
    Returns:
        a_hat : data bits after demodulation
    """
    t = np.arange(start=0,stop=len(r_t))/fs # time base
    x = r_t*(np.cos(2*np.pi*(fc+fd/2)*t+phase)-np.cos(2*np.pi*(fc-fd/2)*t+phase))
    y = np.convolve(x,np.ones(L)) # integrate/sum from 0 to L
    a_hat = (y[L-1::L]>0).astype(int) # sample at every sampling instant and detect
    return a_hat
```

2.12.6 Non-coherent demodulator

A non-coherent FSK demodulator does not care about the initial phase information used at transmitter, hence avoids the need for carrier phase recovery at the receiver. As a result, the non-coherent demodulator is easier to build and therefore offers a better solution for practical applications.

The discrete-time equivalent model for a correlator/square-law based non-coherent demodulator is shown in Figure 2.39. It can be used to demodulate both coherent FSK and non-coherent FSK signal.

2.12 Frequency Shift Keying (FSK)

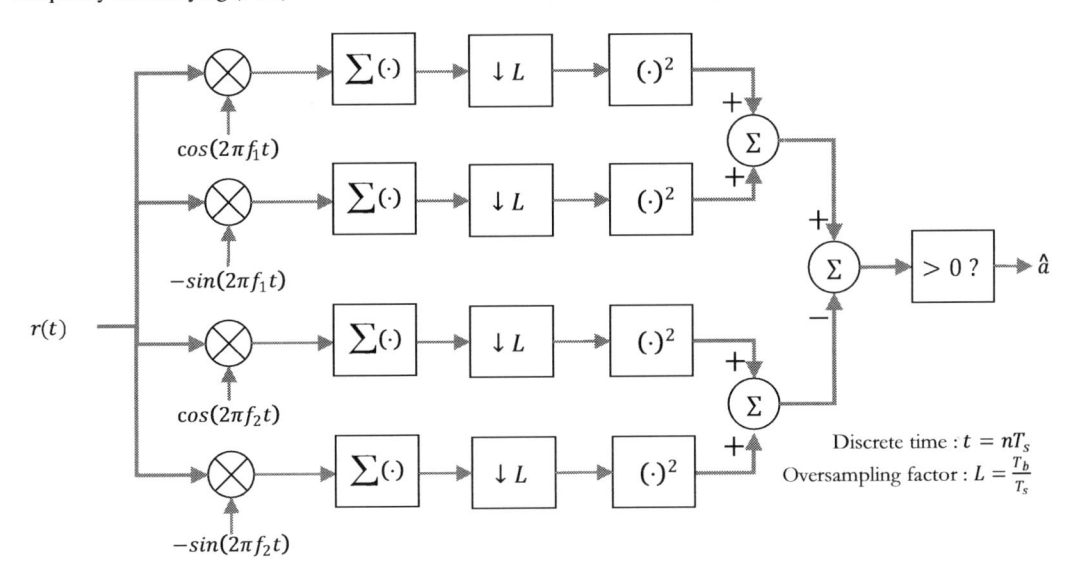

Fig. 2.39: Discrete-time equivalent model for square-law based non-coherent demodulation for coherent/non-coherent FSK signal

The theoretical bit-error probability of non-coherently detected binary FSK over AWGN noise, where two carrier signals are orthogonal and equi-probable, is given by

$$P_b = \frac{1}{2} e^{E_b/(2N_0)} \qquad (2.60)$$

Program 48: DigiCommPy\passband_modulations.py: Square-law based non-coherent demodulator

```
def bfsk_noncoherent_demod(r_t,fc,fd,L,fs):
    """
    Non-coherent demodulation of BFSK modulated signal
    Parameters:
        r_t : BFSK modulated signal at the receiver r(t)
        fc : center frequency of the carrier in Hertz
        fd : frequency separation measured from Fc
        L : number of samples in 1-bit period
        fs : Sampling frequency for discrete-time simulation
    Returns:
        a_hat : data bits after demodulation
    """
    t = np.arange(start=0,stop=len(r_t))/fs # time base
    f1 = (fc+fd/2); f2 = (fc-fd/2)
    #define four basis functions
    p1c = np.cos(2*np.pi*f1*t); p2c = np.cos(2*np.pi*f2*t)
    p1s = -1*np.sin(2*np.pi*f1*t); p2s = -1*np.sin(2*np.pi*f2*t)
    # multiply and integrate from 0 to L
    r1c = np.convolve(r_t*p1c,np.ones(L)); r2c = np.convolve(r_t*p2c,np.ones(L))
    r1s = np.convolve(r_t*p1s,np.ones(L)); r2s = np.convolve(r_t*p2s,np.ones(L))
```

```
    # sample at every sampling instant
    r1c = r1c[L-1::L]; r2c = r2c[L-1::L]
    r1s = r1s[L-1::L]; r2s = r2s[L-1::L]
    # square and add
    x = r1c**2 + r1s**2
    y = r2c**2 + r2s**2
    a_hat=((x-y)>0).astype(int) # compare and decide
    return a_hat
```

2.12.7 Performance simulation

The error rate performances of coherent and non-coherent demodulation of BFSK signal over an AWGN channel is simulated and the results are plotted in Figure 2.40. The simulation code performs BFSK modulation and demodulation by utilizing various functions described in the previous sections: bfsk_mod for generating BFSK modulated signal, bfsk_coherent_demod for coherent demodulation and for non-coherent demodulation the function bfsk_noncoherent_demod is used.

In the code sample below, the FSK type at the transmitter is chosen as coherent. This generates a coherent BFSK signal at the transmitter. The generated signal is then passed through an AWGN channel. The received signal is then independently demodulated using a coherent demodulator and a non-coherent demodulator. For both the cases, the error rate performance is evaluated over a range of E_b/N_0 values.

On the other hand, if the FSK type is chosen as non-coherent in the modulator, the coherent demodulation is not applicable. Therefore, for this case only the performance of the non-coherent demodulator will be plotted.

Program 49: DigiCommPy\chapter_2\bfsk.py: Performance of coherent and non-coherent BFSK

```
#Execute in Python3: exec(open("chapter_2/bfsk.py").read())
import numpy as np #for numerical computing
import matplotlib.pyplot as plt #for plotting functions
from DigiCommPy.passband_modulations import bfsk_mod, bfsk_coherent_demod,
    bfsk_noncoherent_demod
from DigiCommPy.channels import awgn
from scipy.special import erfc

N=100000 # Number of bits to transmit
EbN0dB = np.arange(start=-4,stop = 11, step = 2) # Eb/N0 range in dB for simulation
fc = 400 # center carrier frequency f_c- integral multiple of 1/Tb
fsk_type = 'coherent' # coherent/noncoherent FSK generation at Tx
h = 1 # modulation index
# h should be minimum 0.5 for coherent FSK or multiples of 0.5
# h should be minimum 1 for non-coherent FSK or multiples of 1
L = 40 # oversampling factor
fs = 8*fc # sampling frequency for discrete-time simulation
fd = h/(L/fs) # Frequency separation

BER_coherent = np.zeros(len(EbN0dB)) # BER for coherent BFSK
BER_noncoherent = np.zeros(len(EbN0dB)) # BER for non-coherent BFSK

a = np.random.randint(2, size=N) # uniform random symbols from 0's and 1's
```

2.12 Frequency Shift Keying (FSK)

```
[s_t,phase]=bfsk_mod(a,fc,fd,L,fs,fsk_type) # BFSK modulation

for i, EbN0 in enumerate(EbN0dB):
    r_t = awgn(s_t,EbN0,L) # refer Chapter section 4.1
    
    if fsk_type.lower() == 'coherent':
        # coherent FSK could be demodulated coherently or non-coherently
        a_hat_coherent = bfsk_coherent_demod(r_t,phase,fc,fd,L,fs) # coherent demod
        a_hat_noncoherent = bfsk_noncoherent_demod(r_t,fc,fd,L,fs)#noncoherent demod
        
        BER_coherent[i] = np.sum(a!=a_hat_coherent)/N # BER for coherent case
        BER_noncoherent[i] = np.sum(a!=a_hat_noncoherent)/N # BER for non-coherent
    
    if fsk_type.lower() == 'noncoherent':
        #non-coherent FSK can only non-coherently demodulated
        a_hat_noncoherent = bfsk_noncoherent_demod(r_t,fc,fd,L,fs)#noncoherent demod
        BER_noncoherent[i] = np.sum(a!=a_hat_noncoherent)/N # BER for non-coherent

#Theoretical BERs
theory_coherent = 0.5*erfc(np.sqrt(10**(EbN0dB/10)/2)) # Theory BER - coherent
theory_noncoherent = 0.5*np.exp(-10**(EbN0dB/10)/2) # Theory BER - non-coherent

fig, axs = plt.subplots(1, 1)
if fsk_type.lower() == 'coherent':
    axs.semilogy(EbN0dB,BER_coherent,'k*',label='sim-coherent demod')
    axs.semilogy(EbN0dB,BER_noncoherent,'m*',label='sim-noncoherent demod')
    axs.semilogy(EbN0dB,theory_coherent,'r-',label='theory-coherent demod')
    axs.semilogy(EbN0dB,theory_noncoherent,'b-',label='theory-noncoherent demod')
    axs.set_title('Performance of coherent BFSK modulation')

if fsk_type.lower() == 'noncoherent':
    axs.semilogy(EbN0dB,BER_noncoherent,'m*',label='sim-noncoherent demod')
    axs.semilogy(EbN0dB,theory_noncoherent,'b-',label='theory-noncoherent demod')
    axs.set_title('Performance of noncoherent BFSK modulation')

axs.set_xlabel('$E_b/N_0$ (dB)');axs.set_ylabel('Probability of Bit Error - $P_b$')
axs.legend();fig.show()
```

2.12.8 Power spectral density

Assuming rectangular pulse shaping at the baseband, the power spectral density plot for the BFSK modulated signal is simulated using the windowed Welch spectrum estimation method described in chapter 1 section 1.4 and the result is plotted in Figure 2.41. For this simulation, the center frequency of the carrier is chosen as $f_c = 400Hz$, sampling frequency $f_s = 8 \times f_c$, modulation index $h = 1$ and bit period $T_b = 0.0125$. As expected for this configuration, the PSD plot reveals two spikes at frequencies $f_1 = 440Hz$ and $f_2 = 360Hz$ that are separated from each other by $\Delta f = 80Hz$.

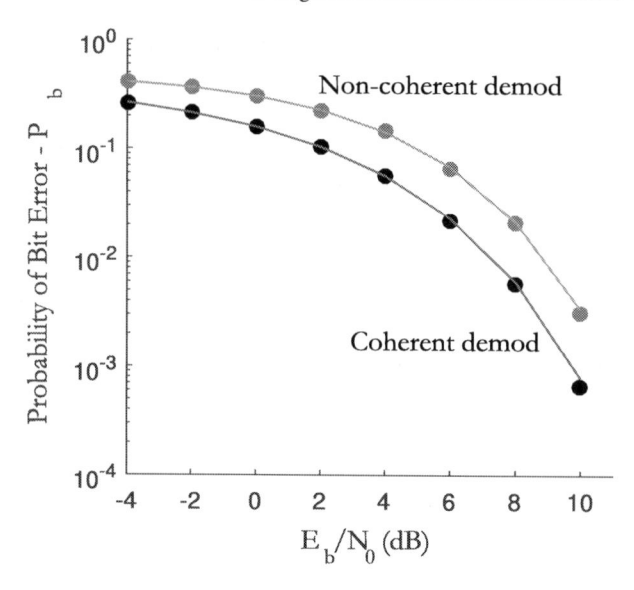

Fig. 2.40: Performance of coherent BFSK when demodulated using coherent and non-coherent techniques

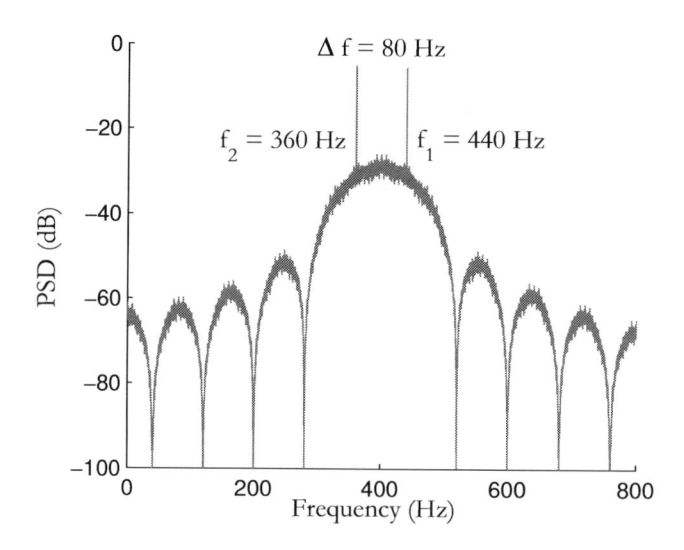

Fig. 2.41: BFSK power spectra

References

1. Lloyd N. Trefethen, David Bau III (1997), *Numerical linear algebra*, Philadelphia: Society for Industrial and Applied Mathematics, ISBN 978-0-89871-361-9, pp.56
2. Park, J. H., Jr., *On binary DPSK detection*,IEEE Trans. Communication, vol. 26, no. 4, April 1978,pp. 484–486
3. Fuqin Xiong,*Digital Modulation Techniques*, Second Edition, Artech House, Inc. ISBN-1580538630
4. S. A. Rhodes,*Effects of hardlimiting on bandlimited transmissions with conventional and offset QPSK modulation*, in Proc. Nat. Telecommun. Conf., Houston, TX, 1972, PP. 20F/1-20F/7
5. S. A. Rhodes,*Effect of noisy phase reference on coherent detection of offset QPSK signals*, IEEE Trans. Commun., vol. COM-22, PP. 1046-1055, Aug. 1974

References

6. S. Pasupathy, *Minimum Shift Keying: A Spectrally Efficient Modulation*, IEEE Communications Magazine, vol. 17, no. 4, pp. 14-22, July 1979
7. Dayan Adionel Guimaraes, *Contributions to the Understanding of the MSK Modulation*, revista telecomunicaes, VOL. 11, NO. 01, MAIO DE 2008.
8. Clay S. Turner, *Raised Cosine and Root Raised Cosine Formulae*, Wireless Systems Engineering, Inc, (May 29, 2007) V1.2
9. Murota, K. and Hirade, K., *GMSK Modulation for Digital Mobile Radio Telephony*, IEEE Transactions on Communications, vol COM-29, No. 7. pp. 1044-1050, July 1981.
10. Marvin K. Simon, *Bandwidth-Efficient Digital Modulation with Application to Deep Space Communications*, JPL Deep Space Communications and Navigation Series, Wiley-Interscience, Hoboken, New Jersey, 2003, ISBN 0-471-44536-3, pp-57.
11. Bridges Paul A, *Digital demodulation techniques for radio channels*, Master thesis, Victoria University of Technology, 1993. http://vuir.vu.edu.au/15572/
12. J. B. Anderson, T. Aulin, and C.-E. Sundberg, *Digital Phase Modulation*, ISBN 978-0306421952, Plenum Press. New York. 1986

Chapter 3
Digital Modulators and Demodulators - Complex Baseband Equivalent Models

Abstract In chapter 2, we saw how waveform level simulations were used to simulate the performance of various modulation techniques. In such simulation models, every cycle of the sinusoidal carrier gets simulated, consequently consuming more memory and time. On the other hand, the complex baseband equivalent models, the subject of this chapter, operates on symbol basis. This approach is much simpler to implement. It offers a great advantage by drastically reducing memory requirements as well as yielding results in a much shorter timespan. The focus of this chapter is on developing simulation codes for implementing various modulations using the complex baseband equivalent approach.

3.1 Introduction

The *passband model* and *equivalent baseband model* are fundamental models for simulating a communication system. In the passband model, also called as *waveform simulation model*, the transmitted signal, channel noise and the received signal are all represented by samples of waveforms. Since every detail of the RF carrier gets simulated, it consumes more memory and time.

In the case of discrete-time equivalent baseband model, only the value of a symbol at the symbol-sampling time instant is considered. Therefore, it consumes less memory and yields results in a very short span of time when compared to the passband models. Such models operate near zero frequency, suppressing the RF carrier and hence the number of samples required for simulation is greatly reduced. They are more suitable for performance analysis simulations. If the behavior of the system is well understood, the model can be simplified further.

3.2 Complex baseband representation of modulated signal

By definition, a passband signal is a signal whose one-sided energy spectrum is centered on non-zero carrier frequency f_c and does not extend to DC. A passband signal or any digitally modulated RF waveform is represented as

$$\tilde{s}(t) = a(t)cos\left[2\pi f_c t + \phi(t)\right] = s_I(t)cos(2\pi f_c t) - s_Q(t)sin(2\pi f_c t) \quad (3.1)$$

where,

$$a(t) = \sqrt{s_I(t)^2 + s_Q(t)^2} \text{ and } \phi(t) = tan^{-1}\left(\frac{s_Q(t)}{s_I(t)}\right) \quad (3.2)$$

Recognizing that the sine and cosine terms in the equation 3.1 are orthogonal components with respect to each other, the signal can be represented in complex form as

$$s(t) = s_I(t) + js_Q(t) \tag{3.3}$$

When represented in this form, the signal $s(t)$ is called the *complex envelope* or the *complex baseband equivalent representation* of the real signal $\tilde{s}(t)$. The components $s_I(t)$ and $s_Q(t)$ are called *inphase* and *quadrature* components respectively. Comparing equations 3.1 and 3.3, it is evident that in the complex baseband equivalent representation, the carrier frequency is suppressed. This greatly reduces both the sampling frequency requirements and the memory needed for simulating the model. Furthermore, equation 3.1, provides a practical way to convert a passband signal to its baseband equivalent and vice-versa [1], as illustrated in Figure 3.1[†]

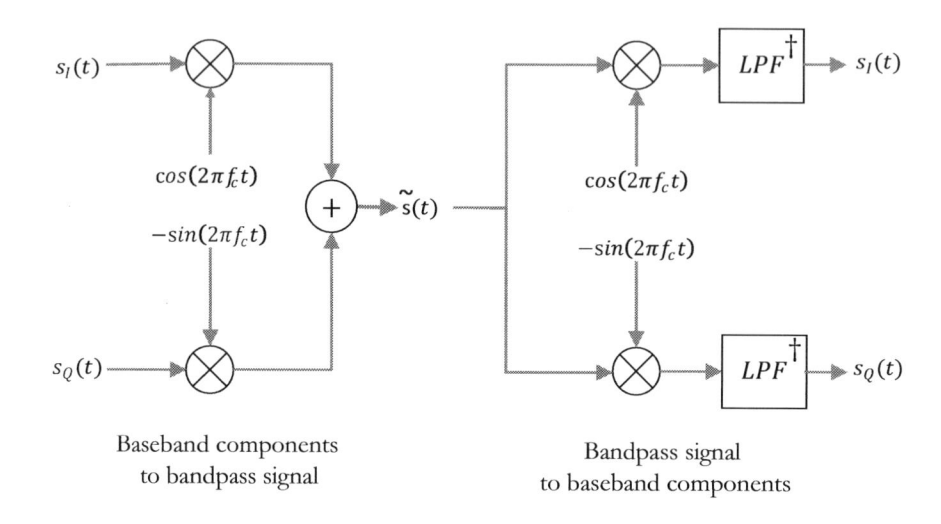

Fig. 3.1: Conversion from baseband to passband and vice-versa

3.3 Complex baseband representation of channel response

In the typical communication system model shown in Figure 3.2(a), the signals represented are real passband signals. The digitally modulated signal $\tilde{s}(t)$ occupies a band-limited spectrum around the carrier frequency (f_c). The channel $\tilde{h}(t)$ is modeled as a *linear time invariant* system which is also band-limited in nature. The effect of the channel on the modulated signal is represented as *linear convolution* (denoted by the $*$ operator). Then, the received signal is given by

$$\tilde{y} = \tilde{s} * \tilde{h} + \tilde{n} \tag{3.4}$$

The corresponding complex baseband equivalent (Figure 3.2(b)) is expressed as

$$(y_I + jy_Q) = (s_I + js_Q) * (h_I + jh_Q) + (n_I + jn_Q) \tag{3.5}$$

[†] LPF = Low Pass Filter

3.4 Implementing complex baseband modems using object oriented programming   123

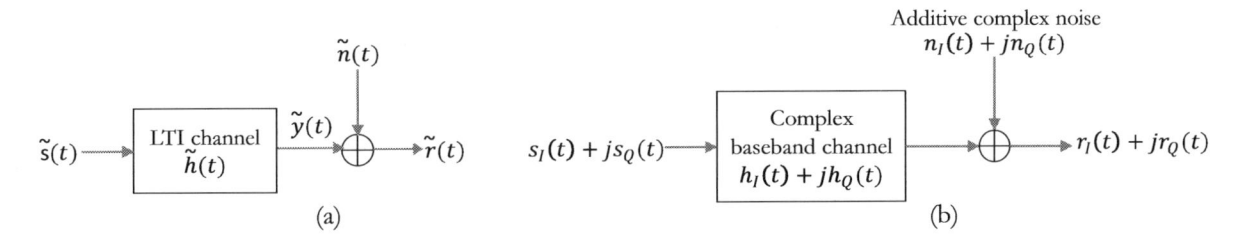

Fig. 3.2: Passband channel model and its baseband equivalent

3.4 Implementing complex baseband modems using object oriented programming

In the following sections, the implementation of modulators and demodulators for the following digital modulation techniques are described:

- Pulse Amplitude Modulation (PAM)
- Pulse Shift Keying Modulation (PSK)
- Quadrature Amplitude Modulation (QAM)
- Orthogonal Frequency Shift Keying Modulation (FSK)

As we can see from the further discussions, all the above mentioned modems have similar structure that can be exploited to write simplified code using *object oriented programming* in Python. We will begin by creating a common *base class* class named Modem and implement the aforementioned modulation techniques as *derived classes*.

The modulate member function of the Modem base class implements the common structure for the aforementioned modulations. The function, simply takes in the input information symbols and selects the corresponding modulated symbols from a reference constellation for the chosen modulation technique. The function is common for PAM, PSK, QAM and FSK modulation techniques. So the derived class implementations for PAM, PSK, QAM and FSK modems, can simply inherit the implementation from the Modem base class, without the need for separate implementation.

The demodulate member function of the Modem base class implements the common structure for the demodulators. The function, takes in the received symbols, performs the optimum detection using an IQ detector and returns the detected symbols. The details of the IQ detector are given in section 3.4.4. The IQ detector demodulation is common for PAM, PSK and QAM modulation techniques. So the derived class implementations for PAM,PSK and QAM modems, can simply inherit the implementation from the Modem base class, without the need for separate implementation. Only for FSK, the detection method needs to overridden.

Program 50: DigiCommPy\modem.py: Defining the Modem base class

```python
import numpy as np
import abc
import matplotlib.pyplot as plt

class Modem:
    __metadata__ = abc.ABCMeta
    # Base class: Modem
    # Attribute definitions:
    #    self.M : number of points in the MPSK constellation
    #    self.name: name of the modem : PSK, QAM, PAM, FSK
    #    self.constellation : reference constellation
```

```
    #     self.coherence : only for 'coherent' or 'noncoherent' FSK
    def __init__(self,M,constellation,name,coherence=None): #constructor
        if (M<2) or ((M & (M -1))!=0): #if M not a power of 2
            raise ValueError('M should be a power of 2')
        if name.lower()=='fsk':
            if (coherence.lower()=='coherent') or (coherence.lower()=='noncoherent'):
                self.coherence = coherence
            else:
                raise ValueError('Coherence must be \'coherent\' or \'noncoherent\'')
        else:
            self.coherence = None
        self.M = M # number of points in the constellation
        self.name = name # name of the modem : PSK, QAM, PAM, FSK
        self.constellation = constellation # reference constellation

    def plotConstellation(self):
        """
        Plot the reference constellation points for the selected modem
        """
        if self.name.lower()=='fsk':
            return 0 # FSK is multi-dimensional difficult to visualize

        fig, axs = plt.subplots(1, 1)
        axs.plot(np.real(self.constellation),np.imag(self.constellation),'o')
        for i in range(0,self.M):
            axs.annotate("{0:0{1}b}".format(i,self.M),
                ↪ (np.real(self.constellation[i]),np.imag(self.constellation[i])))

        axs.set_title('Constellation');
        axs.set_xlabel('I');axs.set_ylabel('Q');fig.show()

    def modulate(self,inputSymbols):
        """
        Modulate a vector of input symbols (numpy array format) using the
        chosen modem. Input symbols take integer values in the range 0 to M-1.
        """
        if isinstance(inputSymbols,list):
            inputSymbols = np.array(inputSymbols)

        if not (0 <= inputSymbols.all() <= self.M-1):
            raise ValueError('inputSymbols values are beyond the range 0 to M-1')

        modulatedVec = self.constellation[inputSymbols]
        return modulatedVec #return modulated vector

    def demodulate(self,receivedSyms):
        """
        Demodulate a vector of received symbols using the chosen modem.
        """
        if isinstance(receivedSyms,list):
```

3.4 Implementing complex baseband modems using object oriented programming 125

```
            receivedSyms = np.array(receivedSyms)

        detectedSyms= self.iqDetector(receivedSyms)
        return detectedSyms

    def iqDetector(self,receivedSyms):
        """
        Optimum Detector for 2-dim. signals (ex: MQAM,MPSK,MPAM) in IQ Plane
        """
        #See section 3.4.4 for definition
```

3.4.1 Pulse Amplitude Modulation (M-PAM) modem

M-PAM modulation is an one dimensional modulation technique that has no quadrature component ($s_Q = 0$). All the information gets encoded in the signal amplitude. A discrete-time baseband model for an M-PAM modulator, transmits a series of information symbols drawn from the set $m \in \{0, 1, ..., M-1\}$ with each transmitted symbol holding k bits of information ($k = log_2(M)$). The information symbols are digitally modulated using M-PAM signaling. The general expression for generating the M-PAM signal constellation set is given by

$$A_m = 2m + 1 - M, \quad m = 0, 1, ..., M-1 \tag{3.6}$$

Here, M denotes the *modulation order* and it defines the number of points in the *ideal reference* constellation. The value of M depends on a parameter k – the number of bits we wish to squeeze in a single M-PAM symbol. For example, if 3 bits ($k = 3$) are squeezed in one transmit symbol, $M = 2^K = 2^3 = 8$, that results in 8-PAM configuration. For, the M-PAM signaling, the constellation points are located at $\pm 1, \pm 3, ..., \pm (M-1)$, as shown in Figure 3.3.

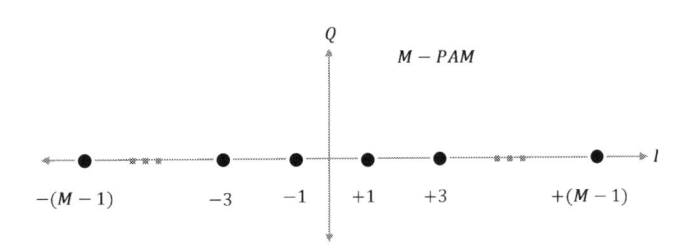

Fig. 3.3: M-PAM constellation

Program 51: DigiCommPy\modem.py: PAM modem - derived class

```python
class PAMModem(Modem):
    # Derived class: PAMModem
    def __init__(self, M):
        m = np.arange(0,M) #all information symbols m={0,1,...,M-1}
        constellation = 2*m+1-M + 1j*0  # reference constellation
        Modem.__init__(self, M, constellation, name='PAM') #set the modem attributes
```

The IQ detection technique based on minimum Euclidean distance metric (described in section 3.4.4) can be leveraged for implementing a coherent detector for M-PAM detection. The following code implements the modem for M-PAM modulation technique by inheriting all the aspects of the Modem base class.

3.4.2 Phase Shift Keying Modulation (M-PSK) modem

In phase shift keying, all the information gets encoded in the phase of the carrier signal. The M-PSK modulator transmits a series of information symbols drawn from the set $m \in \{0, 1, \ldots, M-1\}$. Each transmitted symbol holds k bits of information ($k = log_2(M)$). The information symbols are modulated using M-PSK mapping. The general expression for generating the M-PSK signal set is given by

$$s_m(t) = A cos\left[2\pi f_c t - 2\pi \frac{m}{M}\right], \quad m = 0, 1, \ldots, M-1 \quad (3.7)$$

Here, M denotes the modulation order and it defines the number of constellation points in the reference constellation. The value of M depends on the parameter k – the number of bits we wish to squeeze in a single M-PSK symbol. For example if we wish to squeeze in 3 bits ($k = 3$) in one transmit symbol, then $M = 2^k = 2^3 = 8$ and this results in 8-PSK configuration. $M = 2$ gives BPSK (Binary Phase Shift Keying) configuration. The configuration with $M = 4$ is referred as QPSK (Quadrature Phase Shift Keying). The parameter A is the amplitude scaling factor. Using trigonometric identity, equation 3.7 can be separated into cosine and sine basis functions as follows

$$s_m(t) = A cos\left[2\pi \frac{m}{M}\right] cos(2\pi f_c t) + A sin\left[2\pi \frac{m}{M}\right] sin(2\pi f_c t), \quad m = 0, 1, \ldots, M-1 \quad (3.8)$$

This can be expressed as a combination of in-phase and quadrature phase components on an I-Q plane as

$$s_m(t) = A cos\left[2\pi \frac{m}{M}\right] + jA sin\left[2\pi \frac{m}{M}\right], \quad m = 0, 1, \ldots, M-1 \quad (3.9)$$

Normalizing the amplitude as $A = 1/\sqrt{2}$, the points on the reference constellation will be placed on the unit circle. The MPSK modulator is constructed based on this equation and the ideal constellations for $M = 4, 8$ and 16 PSK modulations are shown in Figure 3.4.

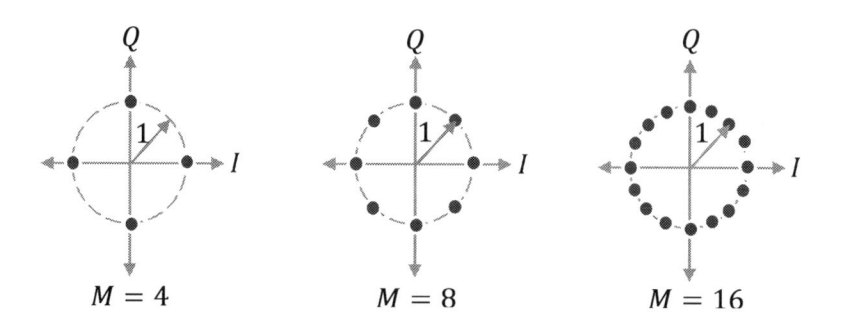

Fig. 3.4: M-PSK constellations

The IQ detection technique based on minimum Euclidean distance metric (described in section 3.4.4) can be leveraged for implementing a coherent detector for MPSK detection.

3.4 Implementing complex baseband modems using object oriented programming 127

Program 52: DigiCommPy\modem.py: PSK modem - derived class

```
class PSKModem(Modem):
    # Derived class: PSKModem
    def __init__(self, M):
        #Generate reference constellation
        m = np.arange(0,M) #all information symbols m={0,1,...,M-1}
        I = 1/np.sqrt(2)*np.cos(m/M*2*np.pi)
        Q = 1/np.sqrt(2)*np.sin(m/M*2*np.pi)
        constellation = I + 1j*Q #reference constellation
        Modem.__init__(self, M, constellation, name='PSK') #set the modem attributes
```

3.4.3 Quadrature Amplitude Modulation (M-QAM) modem

In M-QAM modulations, the information bits are encoded as variations in the amplitude and the phase of the signal. The M-QAM modulator transmits a series of information symbols drawn from the set $m \in \{0,1,..,M-1\}$, with each transmitted symbol holding k bits of information ($k = log2(M)$). To restrict the erroneous receiver decisions to single bit errors, the information symbols are Gray coded. The information symbols are then digitally modulated using a rectangular M-QAM technique, whose signal set is given by

$$s = a + jb \quad where \; a,b \in \{\pm 1, \pm 3, \ldots, \pm (\lceil \sqrt{M} \rceil - 1)\} \tag{3.10}$$

Karnaugh-map walks and Gray codes

In any M-QAM constellation, in order to restrict the erroneous symbol decisions to single bit errors, the adjacent symbols in the transmitter constellation should not differ by more than one bit. This is usually achieved by converting the input symbols to Gray coded symbols and then mapping it to the desired QAM constellation. But this intermediate step can be skipped altogether by using a Look-Up-Table (LUT) approach which properly translates the input symbol to appropriate position in the constellation.

We will exploit the inherent property of Karnaugh Maps to generate the look-up table of dimension $N \times N$ (where $N = \sqrt{M}$) for the Gray coded M-QAM constellation which is rectangular and symmetric ($M = 4,16,64,256,\cdots$). The first step in constructing a QAM constellation is to convert the sequential numbers representing the message symbols to Gray coded format.

If you are familiar with Karnaugh maps (K-Maps), it is easier for you to identify that the K-Maps are constructed based on Gray codes. By the nature of the construction of K-Maps, the address of the adjacent cells differ by only one bit. If we superimpose the given M-QAM constellation on the K-Map and walk through the address of each cell in a certain pattern, it gives the Gray-coded M-QAM constellation.

As mentioned, a walk through the K-Map will produce a sequence of Gray codes. Moreover, if the walk can be looped back to the origin or starting point, it will generate a sequence of cyclic Gray codes. Different walking patterns are possible on K-Maps that generate different sequences of Gray codes. Some of the walks on a 4×4 K-Map are shown in the Figure 3.5. This can be readily extended to any K-Map configuration of higher order.

In walk types 1,3 and 4, the address of the starting point and end point differ by only one bit and the corresponding cells are adjacent to each other. In effect, the walk can be looped to give cyclic Gray codes. But in type 2 walk, the starting cell (0000) and the ending cell (1101) are not adjacent to each other and thus the Gray code generated using this pattern of walk is not cyclic. Thus far, type 1 walk is the simplest. All we have to do is alternate the direction of the walk for every row and read the Gray coded address.

3 Digital Modulators and Demodulators - Complex Baseband Equivalent Models

Walk type 1:
Cyclic Gray code possible

Binary	Dec
0000	0
0001	1
0011	3
0010	2
0110	6
0111	7
0101	5
0100	4
1100	12
1101	13
1111	15
1110	14
1010	10
1011	11
1001	9
1000	8

Walk type 2:
Cyclic Gray code **not** possible

Binary	Dec
0000	0
0001	1
0011	3
0010	2
0110	6
1110	14
1010	10
1011	11
1001	9
1000	8
1100	12
0100	4
0101	5
0111	7
1111	15
1101	13

Walk type 3:
Cyclic Gray code possible

Binary	Dec
0000	0
0100	4
0101	5
0001	1
0011	3
0111	7
0110	6
0010	2
1010	10
1110	14
1111	15
1011	11
1001	9
1101	13
1100	12
1000	8

Walk type 4:
Cyclic Gray code possible

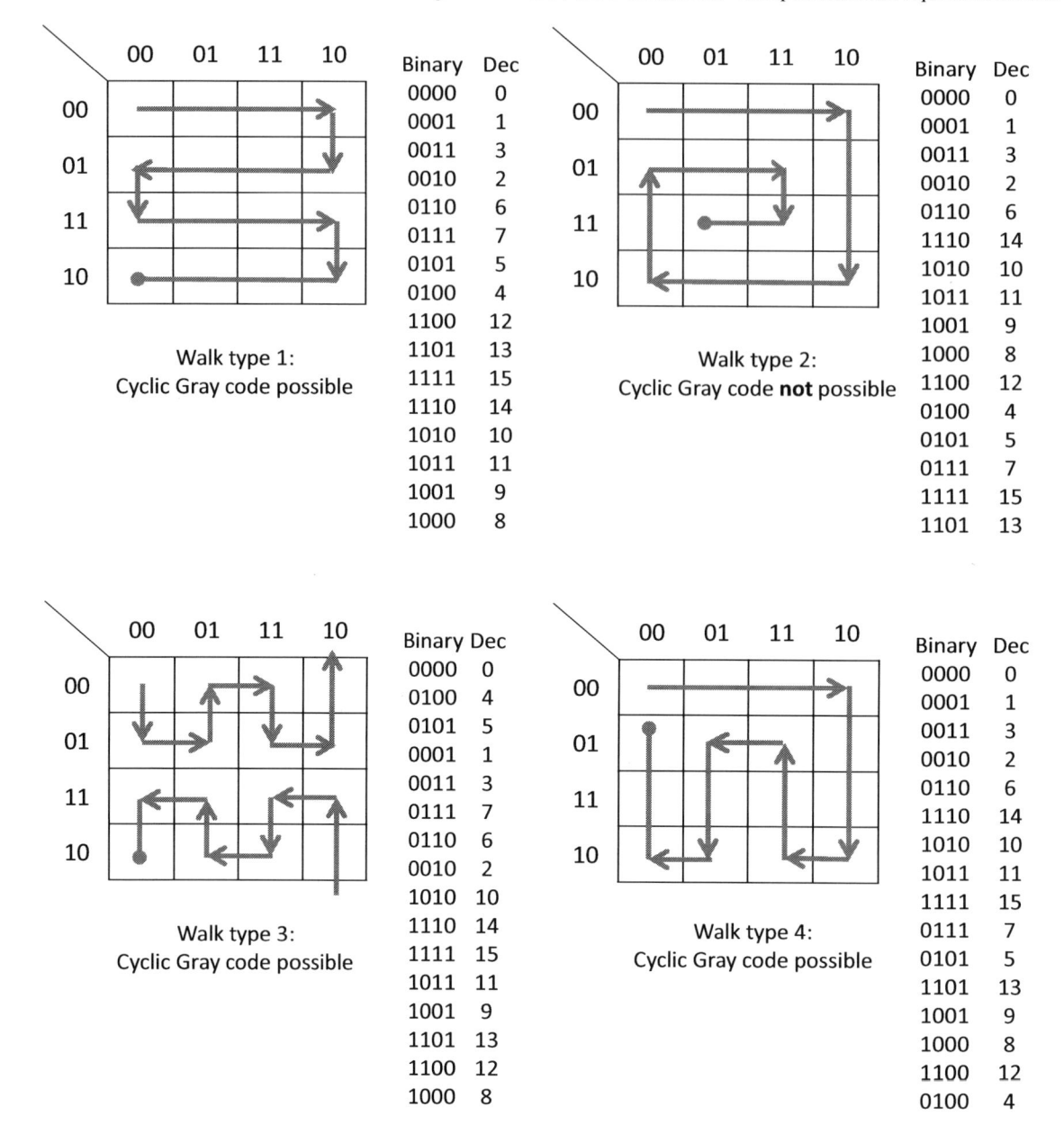

Binary	Dec
0000	0
0001	1
0011	3
0010	2
0110	6
1110	14
1010	10
1011	11
1111	15
0111	7
0101	5
1101	13
1001	9
1000	8
1100	12
0100	4

Fig. 3.5: Karnaugh Map walks and Gray codes

The derived class for QAM modem, defined subsequently, implements the walk type 1 for constructing a M-QAM constellation.

3.4 Implementing complex baseband modems using object oriented programming

Rectangular QAM from PAM constellation

There exist other constellation shapes (like circular, triangular constellations) that are more efficient than the standard rectangular constellation [2]. Symmetric rectangular (square) constellations are the preferred choice of implementation due to their simplicity in implementing modulation and demodulation functions.

Any rectangular QAM constellation is equivalent to superimposing two PAM signals on quadrature carriers. For example, 16-QAM constellation points can be generated from two 4-PAM signals, similarly the 64-QAM constellation points can be generated from two 8-PAM signals. The generic equation to generate PAM signals of dimension D is

$$A_d = 2d + 1 - D, \quad d = 0, 1, \ldots, D - 1 \qquad (3.11)$$

For generating 16-QAM, the dimension D of PAM is set to $D = \sqrt{16} = 4$. Thus for constructing a M-QAM constellation, the PAM dimension is set as $D = \sqrt{M}$.

The M-QAM modulator is implemented as a derived class of the Modem base class. The modulator implementation dynamically generates the M-QAM constellation points based on Karnaugh map Gray code walk type 1. The resulting ideal constellations for Gray coded 16-QAM and 64-QAM are shown in Figure 3.6.

The IQ detection technique based on minimum Euclidean distance metric (described in section 3.4.4) can be leveraged for implementing a coherent detector for MQAM detection.

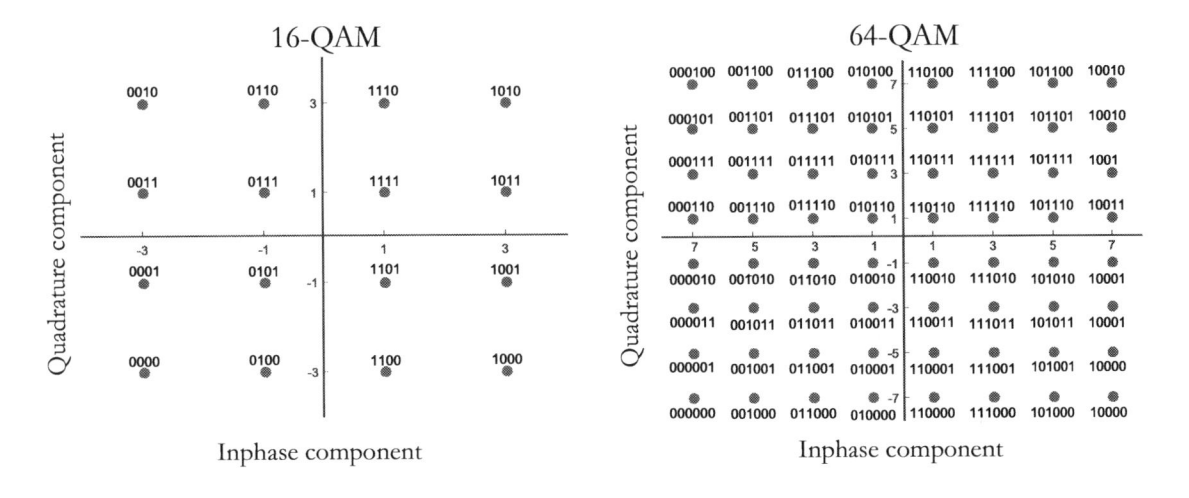

Fig. 3.6: Signal space constellations for 16-QAM and 64-QAM

Program 53: DigiCommPy\modem.py: QAM modem - derived class

```python
class QAMModem(Modem):
    # Derived class: QAMModem
    def __init__(self,M):
        if (M==1) or (np.mod(np.log2(M),2)!=0): # M not a even power of 2
            raise ValueError('Only square MQAM supported. M must be even power of 2')
        
        n = np.arange(0,M) # Sequential address from 0 to M-1 (1xM dimension)
        a = np.asarray([x^(x>>1) for x in n]) #convert linear addresses to Gray code
        D = np.sqrt(M).astype(int) #Dimension of K-Map - N x N matrix
        a = np.reshape(a,(D,D)) # NxN gray coded matrix
        oddRows=np.arange(start = 1, stop = D ,step=2) # identify alternate rows
```

```
        a[oddRows,:] = np.fliplr(a[oddRows,:]) #Flip rows - KMap representation
        nGray=np.reshape(a,(M)) # reshape to 1xM - Gray code walk on KMap

        #Construction of ideal M-QAM constellation from sqrt(M)-PAM
        (x,y)=np.divmod(nGray,D) #element-wise quotient and remainder
        Ax=2*x+1-D # PAM Amplitudes 2d+1-D - real axis
        Ay=2*y+1-D # PAM Amplitudes 2d+1-D - imag axis
        constellation = Ax+1j*Ay
        Modem.__init__(self, M, constellation, name='QAM') #set the modem attributes
```

3.4.4 Optimum detector on IQ plane using minimum Euclidean distance

Two main categories of detection techniques, commonly applied for detecting the digitally modulated data are *coherent detection* and *non-coherent detection*.

In the vector simulation model for the coherent detection, the transmitter and receiver agree on the same reference constellation for modulating and demodulating the information. The modulators described in this section, incorporate the code to generate the reference constellation for the selected modulation type. The same reference constellation should be used if coherent detection is selected as the method of demodulating the received data vector.

On the other hand, in the non-coherent detection, the receiver is oblivious to the reference constellation used at the transmitter. The receiver uses methods like envelope detection to demodulate the data. The IQ detection technique is described here as an example of coherent detection. For an example on the non-coherent detection method, refer Section 3.4.5.2 on non-coherent detection of an MFSK signal vector.

In the IQ detection technique - a type of coherent detection, the first step is to compute the pair-wise Euclidean distance between the given two vectors - reference array and the received symbols corrupted with noise. Each symbol in the received symbol vector (represented on a p-dimensional plane) should be compared with every symbol in the reference array. Next, the symbols, from the reference array, that provide the minimum Euclidean distance are returned.

Let $\mathbf{x} = (x_1, x_2, ..., x_p)$ and $\mathbf{y} = (y_1, y_2, ..., y_p)$ be two points in p-dimensional space. The Euclidean distance between them is given by

$$d(\mathbf{x},\mathbf{y}) = \sqrt{(x_1-y_1)^2 + (x_2-y_2)^2 + ... + (x_p-y_p)^2} \qquad (3.12)$$

The pair-wise Euclidean distance between two sets of vectors, say $\mathbf{x}$ and $\mathbf{y}$, on a p-dimensional space, can be computed using the `cdist` function from the `scipy.spatial.distance` package. The reference symbol that provides the minimum Euclidean distance can be obtained by using the `numpy.argmin` function.

The implemented optimum detector takes in `receivedSymbols` as the input. The vector `receivedSymbols` stands for the sequence of symbols received by the receiver, represented on the complex IQ plane.

The optimum detector computes the pair-wise Euclidean distance of each point in the received vector against every point in the `constellation` vector. The `constellation` vector stands for the ideal constellation points represented in complex IQ plane. It then returns the decoded symbols that provide the minimum Euclidean distance. Since, this optimum detector can be used for IQ modulation techniques like M-PSK, M-QAM, M-PAM and the multidimensional MFSK signaling scheme, it is coded as a member function of the Modem base class, as shown next.

3.4 Implementing complex baseband modems using object oriented programming

Program 54: DigiCommPy\modem.py: Euclidean distance based Optimum IQ Detector

```
class Modem:
    def __init__(self,M,constellation,name,coherence=None): #constructor
        <see section 3.4>

    def plotConstellation(self):
        <see section 3.4>

    def modulate(self,inputSymbols):
        <see section 3.4>

    def demodulate(self,receivedSyms):
        <see section 3.4>

    def iqDetector(self,receivedSyms):
        """
        Optimum Detector for 2-dim. signals (ex: MQAM,MPSK,MPAM) in IQ Plane
        Note: MPAM/BPSK are one dimensional modulations. The same function can be
        applied for these modulations since quadrature is zero (Q=0)

        The function computes the pair-wise Euclidean distance of each point in the
        received vector against every point in the reference constellation. It then
        returns the symbols from the reference constellation that provide the
        minimum Euclidean distance.

        Parameters:
            receivedSyms : received symbol vector of complex form
        Returns:
            detectedSyms : decoded symbols that provide minimum Euclidean distance
        """
        from scipy.spatial.distance import cdist

        # received vector and reference in cartesian form
        XA = np.column_stack((np.real(receivedSyms),np.imag(receivedSyms)))
        XB=np.column_stack((np.real(self.constellation),np.imag(self.constellation)))

        d = cdist(XA,XB,metric='euclidean') #compute pair-wise Euclidean distances
        detectedSyms=np.argmin(d,axis=1)#indices corresponding minimum Euclid. dist.
        return detectedSyms
```

3.4.5 M-ary Frequency Shift Keying modem

M-ary FSK modulation belongs to a broader class of orthogonal modulation, where a transmitted signal corresponding to a source symbol, is drawn from a set of M waveforms whose carrier frequencies are orthogonal to each other. M is the size of the alphabet such that each symbol represents $k = log_2 M$ bits from the information source. When $M = 2$, the format shrinks to BFSK modulation.

3.4.5.1 Modulator for M orthogonal signals

The waveform simulation technique for BFSK was discussed in Section 2.12, where the orthogonality criteria for coherent and non-coherent detection was also derived. The same concept and the simulation technique can be extended for M-ary FSK modulation and detection. However, waveform simulation of M-ary FSK is elaborate and consumes more system memory during run time. A simpler technique to simulate M-ary FSK is to view it in terms of signal space representation. To illustrate this concept we will first see how to represent a BFSK signal using signal space representation and then extend it to represent the broader class of M-ary FSK scheme.

Consider the BFSK modulation scheme, the two waveforms are

$$S_{BFSK}(t) = \begin{cases} s_1(t) = \sqrt{\frac{2E_b}{T_b}} \cos[2\pi f_1 t + \phi_1] & 0 \le t \le T_b, \text{ for binary 1} \\ s_2(t) = \sqrt{\frac{2E_b}{T_b}} \cos[2\pi f_2 t + \phi_2] & 0 \le t \le T_b, \text{ for binary 0} \end{cases}$$

Here, ϕ_1 and ϕ_2 are the unknown random phases of the orthogonal carriers that could take any value from the uniform distribution in the interval $[0, 2\pi)$. If $\phi_1 = \phi_2$, then the modulation is called *coherent* BFSK, otherwise it is called *non-coherent* BFSK. Rewriting equation 3.13,

$$s_1(t) = \sqrt{\frac{2E_b}{T_b}} \cos(2\pi f_1 t)\cos(\phi_1) - \sqrt{\frac{2E_b}{T_b}} \sin(2\pi f_1 t)\sin(\phi_1)$$

$$s_2(t) = \sqrt{\frac{2E_b}{T_b}} \cos(2\pi f_2 t)\cos(\phi_2) - \sqrt{\frac{2E_b}{T_b}} \sin(2\pi f_2 t)\sin(\phi_2) \quad (3.13)$$

Choosing four basis functions,

$$\psi_{1c}(t) = \sqrt{\frac{2}{T_b}}\cos(2\pi f_1 t) \ , \ \psi_{1s}(t) = -\sqrt{\frac{2}{T_b}}\sin(2\pi f_1 t)$$

$$\psi_{2c}(t) = \sqrt{\frac{2}{T_b}}\cos(2\pi f_2 t) \ , \ \psi_{2s}(t) = -\sqrt{\frac{2}{T_b}}\sin(2\pi f_2 t) \quad (3.14)$$

The BFSK waveforms can be expressed in complex notation and the signal set can be conveniently represented as a matrix.

$$\mathbf{s} = \begin{bmatrix} s_1 \\ s_2 \end{bmatrix} = \begin{bmatrix} \sqrt{E_b}\cos(\phi_1) + j\sqrt{E_b}\sin(\phi_1) & 0 \\ 0 & \sqrt{E_b}\cos(\phi_2) + j\sqrt{E_b}\sin(\phi_2) \end{bmatrix}$$

$$= \begin{bmatrix} \sqrt{E_b}e^{j\phi_1} & 0 \\ 0 & \sqrt{E_b}e^{j\phi_2} \end{bmatrix} \quad (3.15)$$

For BFSK modulation, if the source bit $a[k] = 0$, then the vector s_1 will be transmitted and if the source bit $a[k] = 1$, the vector s_2 will be transmitted.

Extending the equation 3.15 to M orthogonal frequencies and when $\phi_1 \ne \phi_2 \ne ... \ne \phi_M$, the *noncoherent* M-ary FSK set is represented as

$$\mathbf{s} = \begin{bmatrix} s_1 \\ s_2 \\ \vdots \\ s_M \end{bmatrix} = \begin{bmatrix} \sqrt{E_b}e^{j\phi_1} & 0 & \cdots & 0 \\ 0 & \sqrt{E_b}e^{j\phi_2} & \cdots & 0 \\ \vdots & \vdots & \ddots & \vdots \\ 0 & 0 & \cdots & \sqrt{E_b}e^{j\phi_M} \end{bmatrix} \quad (3.16)$$

3.4 Implementing complex baseband modems using object oriented programming

For *coherent* M-ary FSK, $\phi_1 = \phi_2 = ... = \phi_M = 0$.

For implementation, the M-FSK modem is implemented as a class *derived* from the Modem base class. The constructor function computes and sets the reference constellation as per equation 3.16.

Program 55: DigiCommPy\modem.py: Defining the the FSK modem derived class

```python
class FSKModem(Modem):
    # Derivied class: FSKModem
    def __init__(self,M,coherence='coherent'):
        if coherence.lower()=='coherent':
            phi= np.zeros(M) # phase=0 for coherent detection
        elif coherence.lower()=='noncoherent':
            phi = 2*np.pi*np.random.rand(M) # M random phases in the (0,2pi)
        else:
            raise ValueError('Coherence must be \'coherent\' or \'noncoherent\'')
        constellation = np.diag(np.exp(1j*phi))
        Modem.__init__(self, M, constellation,
            name='FSK',coherence=coherence.lower()) #set the base modem attributes

    def demodulate(self, receivedSyms,coherence='coherent'):
        <see section 3.4.5.2>
```

3.4.5.2 M-FSK detection

There are two types of detection techniques applicable to MFSK detection: coherent and non-coherent detection. The following function implements the coherent and noncoherent detection of MFSK signal. The theory behind these detection techniques are described next.

Non-coherent detection of MFSK

With reference to equation 3.16, if the MFSK modulated symbol vector s_1 is transmitted at an instance, the received signal vector after passing through an AWGN channel is represented as

$$r = \left(\sqrt{E_b}e^{j\phi_1} + n_1, n_2, \cdots, n_M\right) \quad (3.17)$$

where $n_1, n_2, ..., n_M$ are zero-mean statistically independent Gaussian random variables with equal variance $\sigma^2 = N_0/2$. The detection at receiver can be made simple by employing an *envelope detector*, that chooses the ideal transmitted symbol corresponding to the maximum envelope value. This eliminates the need for the carrier phase reference (the term $e^{j\phi}$) at the receiver and hence the technique is called *non-coherent* detection.

Coherent detection of MFSK

For coherent equal-energy orthogonal signaling (coherent M-FSK), set $\phi = 0$ in equation 3.16. The signal set for coherent signaling is

$$\mathbf{s} = \begin{bmatrix} \mathbf{s}_1 \\ \mathbf{s}_2 \\ \vdots \\ \mathbf{s}_M \end{bmatrix} = \begin{bmatrix} \sqrt{E_b} & 0 & \cdots & 0 \\ 0 & \sqrt{E_b} & \cdots & 0 \\ \vdots & \vdots & \ddots & \vdots \\ 0 & 0 & \cdots & \sqrt{E_b} \end{bmatrix} \tag{3.18}$$

The optimum receiver for AWGN channel computes the pairwise Euclidean distances between each received symbol and the reference signal set as described in section 3.4.4. The symbols from the reference constellation that provides the minimum Euclidean distance is chosen as the detected symbol. Correlation metric can also be used instead of Euclidean distance metric. In such a case, each received symbol is correlated with each of the vectors in the reference array and the vector that provides the maximum correlation can be chosen.

Note that a coherently modulated MFSK signal can be detected using both coherent and non-coherent detection schemes. However, a non-coherently modulated MFSK can only be detected using noncoherent detection. The figure 3.7 captures the simulation model that can be for MFSK modulation and demodulation over an AWGN channel. For a complete performance simulation model, refer the next chapter.

The MFSK demodulator is implemented as a member function of the FSKModem class, as shown next.

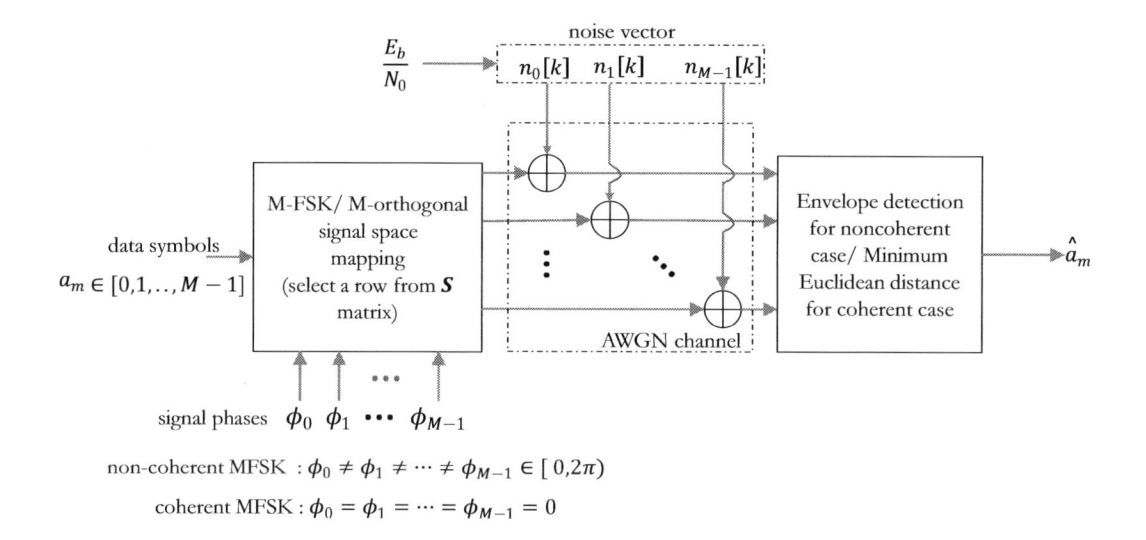

Fig. 3.7: Simulation model for coherent and noncoherently detected MFSK

Program 56: DigiCommPy\modem.py: FSK demodulator

```python
class FSKModem(Modem):
    # Derived class: FSKModem
    def __init__(self,M,coherence='coherent'):
        <see section 3.4.5.1>

    def demodulate(self, receivedSyms,coherence='coherent'):
        #overridden method in Modem class
        if coherence.lower()=='coherent':
            return self.iqDetector(receivedSyms)
        elif coherence.lower()=='noncoherent':
```

```
            return np.argmax(np.abs(receivedSyms),axis=1)
        else:
            raise ValueError('Coherence must be \'coherent\' or \'noncoherent\'')
```

3.5 Instantiation of modems

The intension of this chapter is to show how Python's object oriented approach can be leveraged to implement the different modems. The focus is on the creation of various objects that bind both data and functionality together.

The code definitions for the all the aforementioned modem implementations is written in a single file called modem.py. The complete python code that goes in this file is given here for reference.

Program 57: DigiCommPy\modem.py: Class definitions for various modems

```python
import numpy as np
import abc
import matplotlib.pyplot as plt

class Modem:
    __metadata__ = abc.ABCMeta
    # Base class: Modem
    # Attribute definitions:
    #    self.M : number of points in the MPSK constellation
    #    self.name: name of the modem : PSK, QAM, PAM, FSK
    #    self.constellation : reference constellation
    #    self.coherence : only for 'coherent' or 'noncoherent' FSK
    def __init__(self,M,constellation,name,coherence=None): #constructor
        if (M<2) or ((M & (M -1))!=0): #if M not a power of 2
            raise ValueError('M should be a power of 2')
        if name.lower()=='fsk':
            if (coherence.lower()=='coherent') or (coherence.lower()=='noncoherent'):
                self.coherence = coherence
            else:
                raise ValueError('Coherence must be \'coherent\' or \'noncoherent\'')
        else:
            self.coherence = None
        self.M = M # number of points in the constellation
        self.name = name # name of the modem : PSK, QAM, PAM, FSK
        self.constellation = constellation # reference constellation

    def plotConstellation(self):
        """
        Plot the reference constellation points for the selected modem
        """
        from math import log2
        if self.name.lower()=='fsk':
            return 0 # FSK is multi-dimensional difficult to visualize
```

```python
        fig, axs = plt.subplots(1, 1)
        axs.plot(np.real(self.constellation),np.imag(self.constellation),'o')
        for i in range(0,self.M):
            axs.annotate("{0:0{1}b}".format(i,int(log2(self.M))),
                ↪ (np.real(self.constellation[i]),np.imag(self.constellation[i])))

        axs.set_title('Constellation');
        axs.set_xlabel('I');axs.set_ylabel('Q');fig.show()

    def modulate(self,inputSymbols):
        """
        Modulate a vector of input symbols (numpy array format) using the chosen
            modem. The input symbols take integer values in the range 0 to M-1.
        """
        if isinstance(inputSymbols,list):
            inputSymbols = np.array(inputSymbols)

        if  not (0 <= inputSymbols.all() <= self.M-1):
            raise ValueError('Values for inputSymbols are beyond the range 0 to M-1')

        modulatedVec = self.constellation[inputSymbols]
        return modulatedVec #return modulated vector

    def demodulate(self,receivedSyms):
        """
        Demodulate a vector of received symbols using the chosen modem.
        """
        if isinstance(receivedSyms,list):
            receivedSyms = np.array(receivedSyms)

        detectedSyms= self.iqDetector(receivedSyms)
        return detectedSyms

    def iqDetector(self,receivedSyms):
        """
        Optimum Detector for 2-dim. signals (ex: MQAM,MPSK,MPAM) in IQ Plane
        Note: MPAM/BPSK are one dimensional modulations. The same function can be
        applied for these modulations since quadrature is zero (Q=0)

        The function computes the pair-wise Euclidean distance of each point in the
        received vector against every point in the reference constellation. It then
        returns the symbols from the reference constellation that provide the
        minimum Euclidean distance.

        Parameters:
            receivedSyms : received symbol vector of complex form
        Returns:
            detectedSyms:decoded symbols that provide the minimum Euclidean distance
        ↪
        """
```

3.5 Instantiation of modems

```
            from scipy.spatial.distance import cdist
            # received vector and reference in cartesian form
            XA = np.column_stack((np.real(receivedSyms),np.imag(receivedSyms)))
          XB=np.column_stack((np.real(self.constellation),np.imag(self.constellation)))

            d = cdist(XA,XB,metric='euclidean') #compute pair-wise Euclidean distances
            detectedSyms = np.argmin(d,axis=1)#indices corresponding minimum Euclid.
             ↪ dist.
            return detectedSyms

class PAMModem(Modem):
    # Derived class: PAMModem
    def __init__(self, M):
        m = np.arange(0,M) #all information symbols m={0,1,...,M-1}
        constellation = 2*m+1-M + 1j*0  # reference constellation
        Modem.__init__(self, M, constellation, name='PAM') #set the modem attributes

class PSKModem(Modem):
    # Derived class: PSKModem
    def __init__(self, M):
        #Generate reference constellation
        m = np.arange(0,M) #all information symbols m={0,1,...,M-1}
        I = 1/np.sqrt(2)*np.cos(m/M*2*np.pi)
        Q = 1/np.sqrt(2)*np.sin(m/M*2*np.pi)
        constellation = I + 1j*Q #reference constellation
        Modem.__init__(self, M, constellation, name='PSK') #set the modem attributes

class QAMModem(Modem):
    # Derived class: QAMModem
    def __init__(self,M):
        if (M==1) or (np.mod(np.log2(M),2)!=0): # M not a even power of 2
            raise ValueError('Only square MQAM supported. M must be even power of 2')

        n = np.arange(0,M) # Sequential address from 0 to M-1 (1xM dimension)
        a = np.asarray([x^(x>>1) for x in n]) #convert linear addresses to Gray code
        D = np.sqrt(M).astype(int) #Dimension of K-Map - N x N matrix
        a = np.reshape(a,(D,D)) # NxN gray coded matrix
        oddRows=np.arange(start = 1, stop = D ,step=2) # identify alternate rows
        a[oddRows,:] = np.fliplr(a[oddRows,:]) #Flip rows - KMap representation
        nGray=np.reshape(a,(M)) # reshape to 1xM - Gray code walk on KMap

        #Construction of ideal M-QAM constellation from sqrt(M)-PAM
        (x,y)=np.divmod(nGray,D) #element-wise quotient and remainder
        Ax=2*x+1-D # PAM Amplitudes 2d+1-D - real axis
        Ay=2*y+1-D # PAM Amplitudes 2d+1-D - imag axis
        constellation = Ax+1j*Ay
        Modem.__init__(self, M, constellation, name='QAM') #set the modem attributes

class FSKModem(Modem):
    # Derivied class: FSKModem
```

```python
    def __init__(self,M,coherence='coherent'):
        if coherence.lower()=='coherent':
            phi= np.zeros(M) # phase=0 for coherent detection
        elif coherence.lower()=='noncoherent':
            phi = 2*np.pi*np.random.rand(M) # M random phases in the (0,2pi)
        else:
            raise ValueError('Coherence must be \'coherent\' or \'noncoherent\'')
        constellation = np.diag(np.exp(1j*phi))
        Modem.__init__(self, M, constellation,
            ↪ name='FSK',coherence=coherence.lower()) #set the base modem attributes

    def demodulate(self, receivedSyms,coherence='coherent'):
        #overridden method in Modem class
        if coherence.lower()=='coherent':
            return self.iqDetector(receivedSyms)
        elif coherence.lower()=='noncoherent':
            return np.argmax(np.abs(receivedSyms),axis=1)
        else:
            raise ValueError('Coherence must be \'coherent\' or \'noncoherent\'')
```

The classes defined in the *modem.py* file are just logical entities that define the behavior (methods) and properties (attributes) of different modems. The required modem class need to be instantiated before it could be utilized in further programming. When the modem classes are instantiated, a modem object, which is an entity with specific characteristic or function, is created. The following code demonstrates the instantiation of a 16-PSK modem object and how its member functions can be invoked to perform modulation and demodulation.

```
>> import sys
>> sys.path.append('path_where_digicommpy_resides') #search path
>> from DigiCommPy.modem import PSKModem #import the PSKModem class from modem.py
>> M = 16 #16 points in the constellation
>> pskModem = PSKModem(M) #create a 16-PSK modem object
>> pskModem.plotConstellation() #plot ideal constellation for this modem
>> import numpy as np # for numerical computing
>> nSym = 10 #10 symbols as input to PSK modem
>> inputSyms = np.random.randint(low=0, high = M, size=nSym) # uniform random
    ↪ symbols from 0 to M-1
array([10, 14,  1,  0,  0,  1, 10,  0, 14,  5])
>> modulatedSyms = pskModem.modulate(inputSyms) #modulate
array([-0.5-0.5j, 0.5-0.5j , 0.65+0.27j, 0.707+0.j, 0.707+0.j, 0.65+0.27j,
    ↪ -0.5-0.5j, 0.707+0.j, 0.5-0.5j, -0.27+0.65j])
>> detectedSyms = pskModem.demodulate(modulatedSyms) #demodulate
array([10, 14,  1,  0,  0,  1, 10,  0, 14,  5], dtype=int64)
```

References

1. Proakis J.G, *Digital Communications*, fifth edition, New York, McGraw–Hill, 2008
2. Jiaqi Zhao et al., *Investigation on performance of special-shaped 8-quadrature amplitude modulation constellations applied in visible light communication*, Photon. Res. 4, 249-256 (2016)

Chapter 4
Performance of Digital Modulations over Wireless Channels

Abstract The focus of this chapter is to develop code for performance simulation of various digital modulation techniques over AWGN and flat fading channels. Complex baseband models, developed in chapter 3, are best suited for performance simulation since they consume less computer memory and yield results quickly. Theoretical performance symbol error rates for AWGN and flat fading channels are also provided in this chapter and that can be used as a benchmark to verify the implemented simulation models

4.1 AWGN channel

In this section, the relationship between SNR-per-bit (E_b/N_0) and SNR-per-symbol (E_s/N_0) are defined with respect to M-ary signaling schemes. Then the complex baseband model for an AWGN channel is discussed, followed by the theoretical error rates of various modulations over the *additive white Gaussian noise* (AWGN) channel. Finally, the complex baseband models for digital modulators and detectors developed in chapter 3, are incorporated to build a complete communication system model.

4.1.1 Signal to noise ratio (SNR) definitions

Assuming a channel of bandwidth B, received signal power P_r and the power spectral density (PSD) of noise $N_0/2$, the signal to noise ratio (SNR) is given by

$$\gamma = \frac{P_r}{N_0 B} \tag{4.1}$$

Let a signal's energy-per-bit is denoted as E_b and the energy-per-symbol as E_s, then $\gamma_b = E_b/N_0$ and $\gamma_s = E_s/N_0$ are the SNR-per-bit and the SNR-per-symbol respectively. For uncoded M-ary signaling scheme with $k = log_2(M)$ bits per symbol, the signal energy per modulated symbol is given by

$$E_s = k.E_b \tag{4.2}$$

The SNR per symbol is given by

$$\gamma_s = \frac{E_s}{N_0} = k.\frac{E_b}{N_0} = k.\gamma_b \tag{4.3}$$

4.1.2 AWGN channel model

In order to simulate a specific SNR point in performance simulations, the modulated signal from the transmitter needs to be added with random noise of specific strength. The strength of the generated noise depends on the desired SNR level which usually is an input in such simulations. In practice, SNRs are specified in *dB*. Given a specific SNR point for simulation, let's see how we can simulate an AWGN channel that adds correct level of white noise to the transmitted symbols.

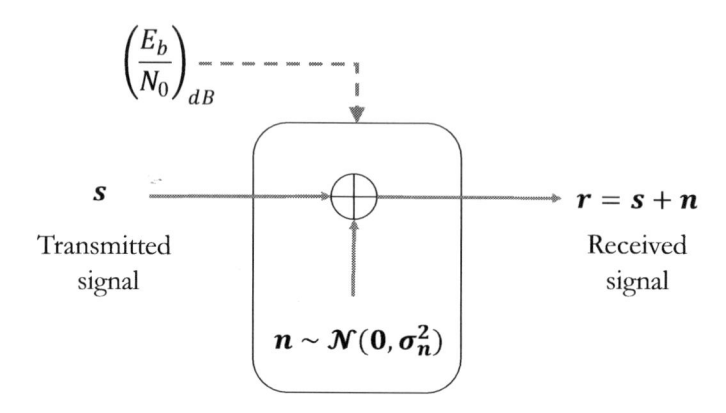

Fig. 4.1: AWGN noise model - computes and adds white Gaussian noise vector for a given SNR value

Consider the AWGN channel model given in Figure 4.1. Given a specific SNR point to simulate, we wish to generate a white Gaussian noise vector $\mathcal{N}(0, \sigma^2)$ of appropriate strength and add it to the incoming signal. The method described can be applied for both waveform simulations and the complex baseband simulations. In following text, the term SNR (γ) refers to $\gamma_b = E_b/N_0$ when the modulation is of binary type (example: BPSK). For multilevel modulations such as QPSK and MQAM, the term SNR refers to $\gamma_s = E_s/N_0$.

1. Assume, **s** is a vector that represents the transmitted signal. We wish to generate a vector **r** that represents the signal after passing through the AWGN channel. The amount of noise added by the AWGN channel is controlled by the given SNR - γ.
2. For waveform simulation (see chapter 2) model, let the given oversampling ratio is denoted as L. On the other hand, if you are using the complex baseband models (see chapter 3), set $L = 1$.
3. Let N denotes the length of the vector **s**. The signal power for the vector **s** can be measured as,

$$P = L \times \frac{1}{N} \sum_{i=0}^{N-1} |s_i|^2 \qquad (4.4)$$

4. The required power spectral density of the noise vector **n** is computed as

$$N_0 = \frac{signal\ power}{signal-to-noise-ratio} = \frac{P}{\gamma} \qquad (4.5)$$

5. Assuming complex IQ plane for all the digital modulations, the required noise variance (noise power) for generating Gaussian random noise is given by

$$\sigma^2 = \frac{N_0}{2} \qquad (4.6)$$

4.1 AWGN channel

6. Finally, generate the Gaussian random noise vector **n** of length N whose samples are drawn from a Gaussian distribution with mean set to zero and the standard deviation as computed in equation 4.6.

$$\mathbf{n} = \begin{cases} \sigma \times \mathcal{N}_N(0,1) & \text{if } \mathbf{s} \text{ is real} \\ \sigma \times [\mathcal{N}_N(0,1) + j * \mathcal{N}_N(0,1)] & \text{if } \mathbf{s} \text{ is complex} \end{cases} \quad (4.7)$$

7. Finally add the generated noise vector to the signal **s**

$$\mathbf{r} = \mathbf{s} + \mathbf{n} \quad (4.8)$$

The following custom function written in Python, can be used for adding AWGN noise to an incoming signal. It can be used in waveform simulation as well as complex baseband simulation models.

Program 58: DigiCommPy\channels.py: A custom function to add AWGN noise to a signal vector

```python
from numpy import sum,isrealobj,sqrt
from numpy.random import standard_normal

def awgn(s,SNRdB,L=1):
    """
    AWGN channel

    Add AWGN noise to input signal. The function adds AWGN noise vector to signal
    's' to generate a resulting signal vector 'r' of specified SNR in dB. It also
    returns the noise vector 'n' that is added to the signal 's' and the power
    spectral density N0 of noise added

    Parameters:
        s : input/transmitted signal vector
        SNRdB : desired signal to noise ratio (expressed in dB)
            for the received signal
        L : oversampling factor (applicable for waveform simulation)
            default L = 1.
    Returns:
        r : received signal vector (r=s+n)
    """
    gamma = 10**(SNRdB/10) #SNR to linear scale

    if s.ndim==1:# if s is single dimensional vector
        P=L*sum(abs(s)**2)/len(s) #Actual power in the vector
    else: # multi-dimensional signals like MFSK
        P=L*sum(sum(abs(s)**2))/len(s) # if s is a matrix [MxN]

    N0=P/gamma # Find the noise spectral density
    if isrealobj(s):# check if input is real/complex object type
        n = sqrt(N0/2)*standard_normal(s.shape) # computed noise
    else:
        n = sqrt(N0/2)*(standard_normal(s.shape)+1j*standard_normal(s.shape))
    r = s + n # received signal
    return r
```

4.1.3 Theoretical symbol error rates

Denoting the *symbol error rate* (SER) as P_s, SNR-per-bit as $\gamma_b = E_b/N_0$ and SNR-per-symbol as $\gamma_s = E_s/N_0$, the symbol error rates for various modulation schemes over AWGN channel are listed in Table 4.1.

Modulation	Symbol error rate (P_s)
MPAM	$2\left(1 - \dfrac{1}{M}\right) Q\left(\sqrt{\dfrac{6}{M^2-1}\gamma_s}\right)$
BPSK	$Q\left(\sqrt{2\gamma_b}\right)$
QPSK	$2Q\left(\sqrt{2\gamma_b}\right) - Q^2\left(\sqrt{2\gamma_b}\right)$
MPSK ($M > 4$)	$2Q\left[\sin\left(\dfrac{\pi}{M}\right)\sqrt{2\gamma_s}\right]$
MQAM	$1 - \left[1 - 2\left(1 - \dfrac{1}{\sqrt{M}}\right) Q\left(\sqrt{\dfrac{3\gamma_s}{(M-1)}}\right)\right]^2$
MFSK (noncoherent)	$\sum_{i=1}^{M-1} \dfrac{(-1)^{i+1}}{i+1} \binom{M-1}{i} exp\left(-\dfrac{i}{i+1}\gamma_s\right)$
MFSK (coherent)	$1 - \int_{-\infty}^{\infty} \left[Q\left(-q - \sqrt{2\gamma_s}\right)\right]^{M-1} \dfrac{1}{\sqrt{2\pi}} exp\left(-\dfrac{q^2}{2}\right) dq$

Table 4.1: Theoretical symbol error rates for various digital modulation schemes

The theoretical symbol error rates are coded as a reusable function (shown next). In this implementation, $erfc$ function is used instead of the Q function shown in the Table 4.1. The following equation describes the relationship between the $erfc$ function and the Q function.

$$Q(x) = \frac{1}{2} erfc\left(\frac{x}{\sqrt{2}}\right) \qquad (4.9)$$

Program 59: DigiCommPy\errorRates.py: Theoretical SERs for various modulation over AWGN

```python
import numpy as np
from numpy import log2,sqrt,sin,pi,exp
from scipy.special import erfc
from scipy.integrate import quad

def ser_awgn(EbN0dBs,mod_type=None,M=0,coherence=None):
    """
    Theoretical Symbol Error Rates for various modulations over AWGN
    Parameters:
        EbN0dBs : list of SNR per bit values in dB scale
        mod_type : 'PSK','QAM','PAM','FSK'
        M : Modulation level for the chosen modulation.
            For PSK,PAM,FSK M can be any power of 2.
```

4.1 AWGN channel

```
                For QAM M must be even power of 2 (square QAM only)
            coherence : 'coherent' for coherent FSK detection
                        'noncoherent' for noncoherent FSK detection
                    This parameter is only applicable for FSK modulation
    Returns:
        SERs = Symbol Error Rates
    """
    if mod_type==None:
        raise ValueError('Invalid value for mod_type')
    if (M<2) or ((M & (M -1))!=0): #if M not a power of 2
        raise ValueError('M should be a power of 2')

    func_dict = {'psk': psk_awgn,'qam':qam_awgn,'pam':pam_awgn,'fsk':fsk_awgn}

    gamma_s = log2(M)*(10**(EbN0dBs/10))
    if mod_type.lower()=='fsk': #call appropriate function
        return func_dict[mod_type.lower()](M,gamma_s,coherence)
    else:
        return func_dict[mod_type.lower()](M,gamma_s) #call appropriate function

def psk_awgn(M,gamma_s):
    gamma_b = gamma_s/log2(M)
    if (M==2):
        SERs = 0.5*erfc(sqrt(gamma_b))
    elif M==4:
        Q = 0.5*erfc(sqrt(gamma_b))
        SERs = 2*Q-Q**2
    else:
        SERs = erfc(sqrt(gamma_s)*sin(pi/M))
    return SERs

def qam_awgn(M,gamma_s):
    if (M==1) or (np.mod(np.log2(M),2)!=0): # M not a even power of 2
        raise ValueError('Only square MQAM supported. M must be even power of 2')
    SERs = 1-(1-(1-1/sqrt(M))*erfc(sqrt(3/2*gamma_s/(M-1))))**2
    return SERs

def pam_awgn(M,gamma_s):
    SERs=2*(1-1/M)*0.5*erfc(sqrt(3*gamma_s/(M**2-1)))
    return SERs

def integrand(q,gamma_s,M):
    return (0.5*erfc((-q-np.sqrt(2*gamma_s))/np.sqrt(2)))**(M-1)\
        *1/np.sqrt(2*pi)*np.exp(-(q**2)/2)

def fsk_awgn(M_val,gamma_s_vals,coherence):
    SERs = np.zeros(len(gamma_s_vals))
    if coherence.lower()=='coherent':
        for j,gamma_s in enumerate(gamma_s_vals):
            (y,_) =  quad(integrand,-np.inf,np.inf,(gamma_s,M_val))
```

```
            SERs[j] = 1- y
    elif coherence.lower()=='noncoherent':
        #use SymPy - symbolic mathematics for evaluating the SER equations
        from sympy import symbols,Symbol,Sum,exp,binomial,erfc,integrate,oo,sqrt
        M,i = symbols('M i', integer=True, positive=True)
        gamma = Symbol('gamma')
        s = Sum((-1)**(i+1)/(i+1)*binomial(M-1,i)*exp(-i/(i+1)*gamma),(i,1,M-1))
        for j,gamma_s in enumerate(gamma_s_vals):
            #evaluate the expression with values for M and gamma_s
            SERs[j] = s.evalf(subs={M:M_val,gamma:gamma_s})
    else:
        raise ValueError('For FSK coherence must be \'coherent\' or \'noncoherent\'')
    return SERs
```

4.1.4 Unified simulation model for performance simulation

In chapter 3, the code implementation for complex baseband models for various digital modulators and demodulator are given. The AWGN channel model is given section 4.1.2. Using these models, we can create a unified simulation code for simulating the performance of various modulation techniques over AWGN channel.

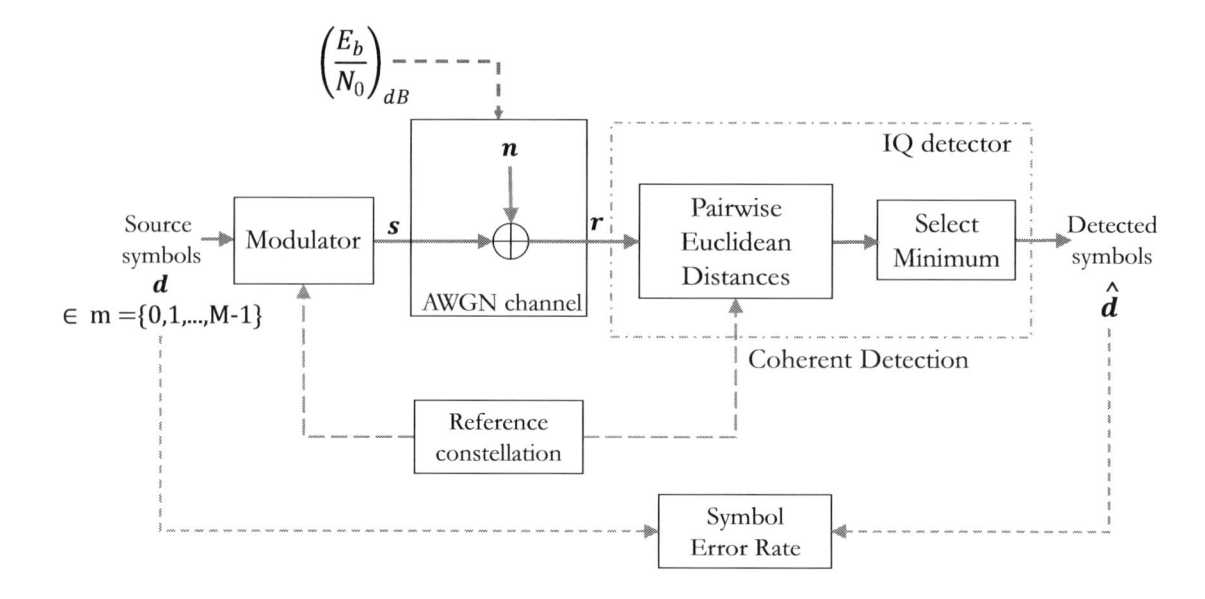

Fig. 4.2: Performance simulation model illustrated for a receiver employing coherent detection

The complete simulation model for performance simulation over AWGN channel is given in Figure 4.2. The figure is illustrated for a *coherent* communication system model (applicable for MPSK/MQAM/M-PAM/MFSK modulations)

Figure 4.3 illustrates the simulation model for multi-level signaling like MFSK modulation. As discussed in section 3.4.5 of chapter 3, a coherently modulated MFSK signal can be detected using both coherent de-

4.1 AWGN channel

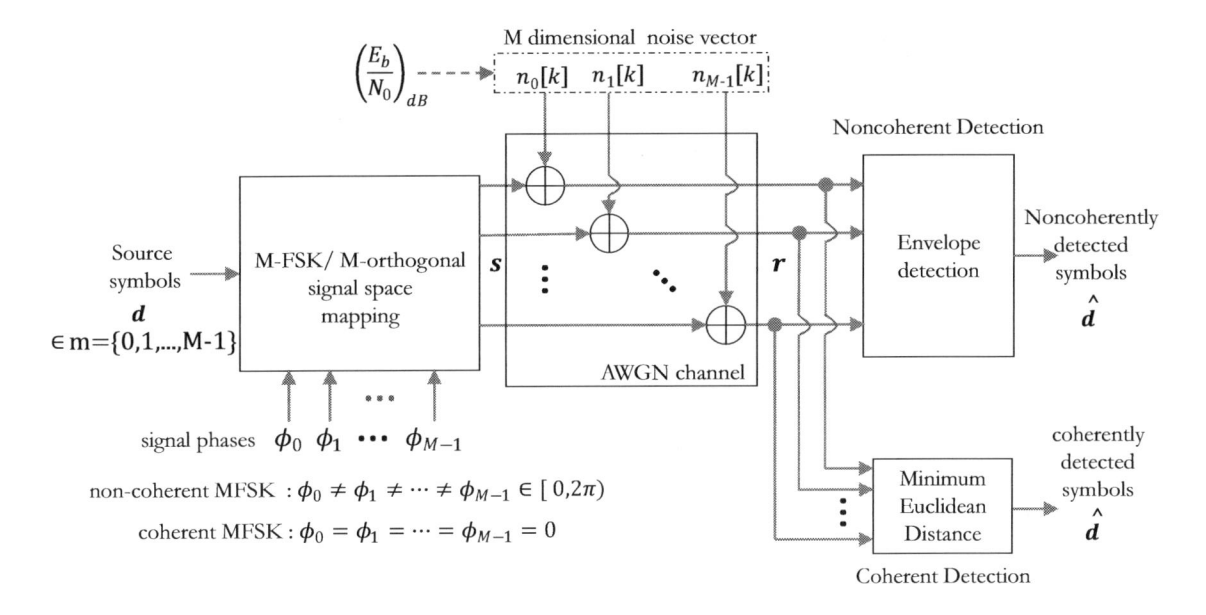

Fig. 4.3: Performance simulation model illustrated for MFSK modulation over AWGN channel

tection and noncoherent detection technique. However, if the MFSK modulator employs noncoherent signal generation, the receiver can only employ noncoherent detection technique.

The Python code implementing the aforementioned simulation model is given next. Here, an unified approach is employed to simulate the performance of any of the given modulation technique - MPSK, MQAM, MPAM or MFSK. To simulate the performance for a given modulation, the user just needs to set the mod_type variable to 'PAM' or 'PSK' or 'QAM' or 'FSK'. If FSK is chosen a the modulation type, the user needs to specify if the receiver employs 'coherent' or 'noncoherent' detection technique, by setting the coherence variable in the code.

The simulation code will automatically choose the selected modulation type, performs Monte Carlo simulation, computes symbol error rates and plots them against the theoretical symbol error rates. The simulated performance results obtained for various modulations are shown in the Figure 4.4.

Program 60: DigiCommPy\chapter_4\awgnPerformance.py: Performance of modulations in AWGN

```
import numpy as np # for numerical computing
import matplotlib.pyplot as plt # for plotting functions
from matplotlib import cm # colormap for color palette
from scipy.special import erfc
from DigiCommPy.modem import PSKModem,QAMModem,PAMModem,FSKModem
from DigiCommPy.channels import awgn
from DigiCommPy.errorRates import ser_awgn

#---------Input Fields------------------
nSym = 10**6 # Number of symbols to transmit
EbN0dBs = np.arange(start=-4,stop = 12, step = 2) # Eb/N0 range in dB for simulation
mod_type = 'FSK' # Set 'PSK' or 'QAM' or 'PAM' or 'FSK'
arrayOfM = [2,4,8,16,32] # array of M values to simulate
#arrayOfM=[4,16,64,256] # uncomment this line if MOD_TYPE='QAM'
```

```python
    coherence = 'coherent' #'coherent'/'noncoherent'-only for FSK

    modem_dict = {'psk': PSKModem,'qam':QAMModem,'pam':PAMModem,'fsk':FSKModem}
    colors = plt.cm.jet(np.linspace(0,1,len(arrayOfM))) # colormap
    fig, ax = plt.subplots(nrows=1,ncols = 1)

    for i, M in enumerate(arrayOfM):
        #-----Initialization of various parameters----
        k=np.log2(M)
        EsN0dBs = 10*np.log10(k)+EbN0dBs # EsN0dB calculation
        SER_sim = np.zeros(len(EbN0dBs)) # simulated Symbol error rates
        inputSyms = np.random.randint(low=0, high = M, size=nSym)
        # uniform random symbols from 0 to M-1

        if mod_type.lower()=='fsk':
            modem=modem_dict[mod_type.lower()](M,coherence)#choose modem from dictionary
        else: #for all other modulations
            modem = modem_dict[mod_type.lower()](M)#choose modem from dictionary
        modulatedSyms = modem.modulate(inputSyms) #modulate

        for j,EsN0dB in enumerate(EsN0dBs):
            receivedSyms = awgn(modulatedSyms,EsN0dB) #add awgn noise

            if mod_type.lower()=='fsk': #demodulate (Refer Chapter 3)
                detectedSyms = modem.demodulate(receivedSyms,coherence)
            else: #demodulate (Refer Chapter 3)
                detectedSyms = modem.demodulate(receivedSyms)

            SER_sim[j] = np.sum(detectedSyms != inputSyms)/nSym

        SER_theory = ser_awgn(EbN0dBs,mod_type,M,coherence) #theory SER
        ax.semilogy(EbN0dBs,SER_sim,color =
            colors[i],marker='o',linestyle='',label='Sim '+str(M)+'-'+mod_type.upper())
        ax.semilogy(EbN0dBs,SER_theory,color = colors[i],linestyle='-',label='Theory
            '+str(M)+'-'+mod_type.upper())
    ax.set_xlabel('Eb/N0(dB)');ax.set_ylabel('SER ($P_s$)')
    ax.set_title('Probability of Symbol Error for M-'+str(mod_type)+' over AWGN')
    ax.legend();fig.show()
```

4.2 Fading channels

With respect to the frequency domain characteristics, the fading channels can be classified into *frequency selective* and *flat fading* channel. The channel fading can be modeled with different statistics like Rayleigh, Rician, Nakagami fading. The simulation models, in this section, are focused on obtaining the simulated performance of various modulations over Rayleigh flat fading and Rician flat fading channels. Modeling of frequency selective fading channel is discussed in reference [1].

4.2 Fading channels

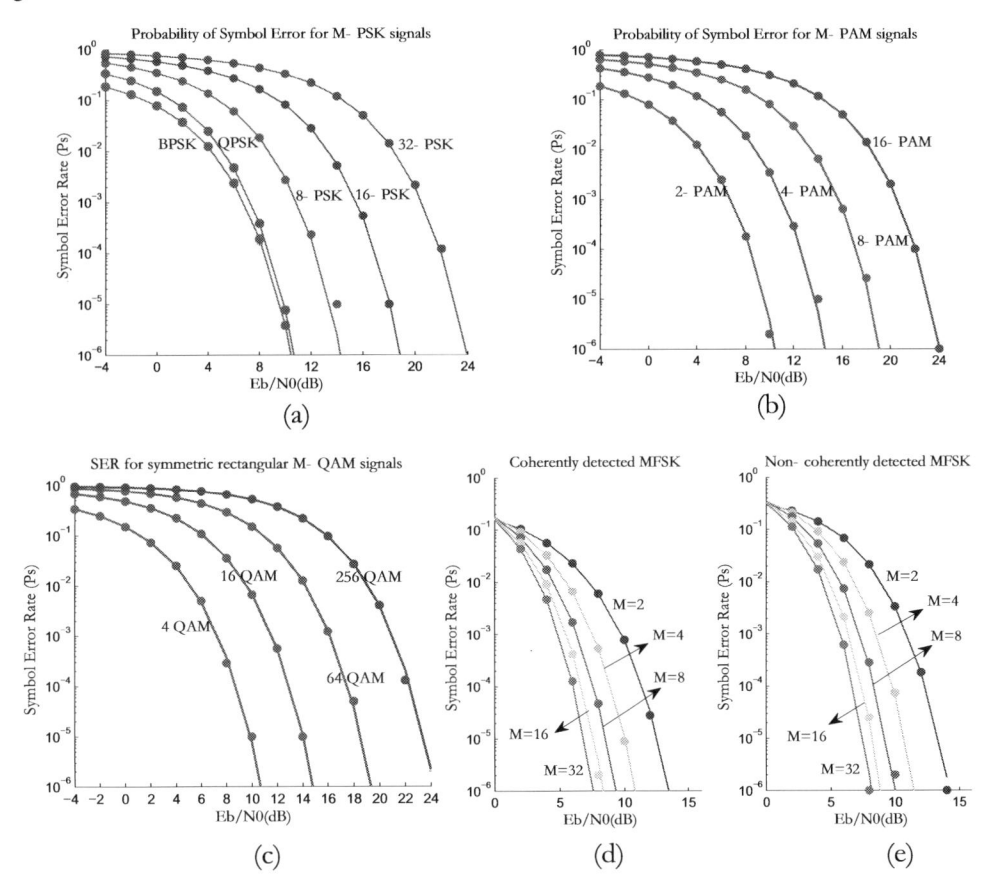

Fig. 4.4: Performance of modulations over AWGN channel : (a) MPSK , (b) MPAM, (c) MQAM, (d) coherently detected MFSK (e) noncoherently detected MFSK

4.2.1 Linear time invariant channel model and FIR filters

The most significant feature of a real world channel is that the channel does not immediately respond to the input. Physically, this indicates some sort of inertia built into the channel/medium, that takes some time to respond. As a consequence, it may introduce distortion effects like *inter-symbol interference* (ISI) at the channel output. Such effects are best studied with the *linear time invariant* (LTI) channel model, given in Figure 4.5.

In this model, the channel response to any input depends only on the *channel impulse response* (CIR) function of the channel. The CIR is usually defined for finite length L as $\mathbf{h} = [h_0, h_1, h_2, ..., h_{L-1}]$ where h_0 is the CIR at symbol sampling instant $0 T_{sym}$ and h_{L-1} is the CIR at symbol sampling instant $(L-1)T_{sym}$. Such a channel can be modeled as a *tapped delay line* (TDL) filter, otherwise called *finite impulse response* (FIR) filter. Here, we only consider the CIR at symbol sampling instances. It is well known that the output of such a channel ($\mathbf{r}$) is given as the *linear convolution* of the input symbols ($\mathbf{s}$) and the CIR ($\mathbf{h}$) at symbol sampling instances. In addition, channel noise in the form of AWGN can also be included the model. Therefore, the resulting vector of from the entire channel model is given as

$$\mathbf{r} = \mathbf{h} * \mathbf{s} + \mathbf{n} \tag{4.10}$$

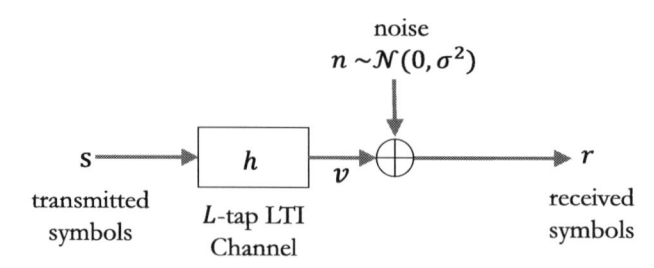

Fig. 4.5: Linear time invariant (LTI) channel model

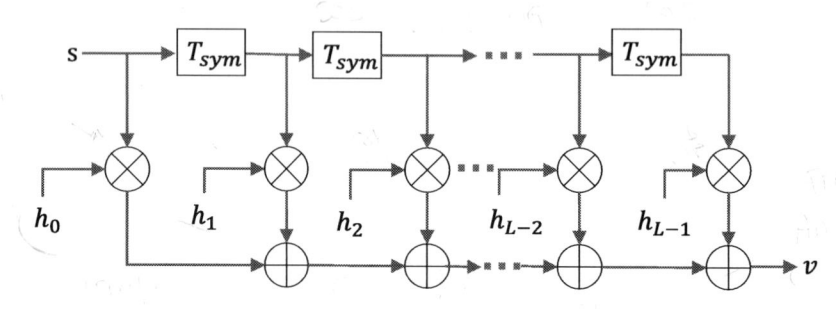

Fig. 4.6: Linear time invariant (LTI) channel viewed as a tapped delay line (TDL) filter

4.2.2 Simulation model for detection in flat fading channel

A flat-fading (also called as *frequency-non-selective*) channel is modeled with a single tap ($L = 1$) FIR filter with the tap weights drawn from distributions like Rayleigh, Rician or Nakagami distributions. We will assume *block fading*, which implies that the fading process is approximately constant for a given transmission interval. For block fading, the random tap coefficient $h = h[0]$ is a *complex* random variable (not random processes) and for each channel realization, a new set of random values are drawn from Rayleigh or Rician or Nakagami fading according to the type of fading desired.

Simulation models for modulation and detection over a fading channel is shown in Figure 4.7. For a flat fading channel, the output of the channel can be expressed simply as the product of time varying channel response and the input signal. Thus, equation 4.10 can be simplified as follows for the flat fading channel.

$$\mathbf{r} = h\mathbf{s} + \mathbf{n} \qquad (4.11)$$

Since the channel and noise are modeled as a complex vectors, the detection of **s** from the received signal is an estimation problem in the complex vector space.

Assuming perfect channel knowledge at the receiver and coherent detection, the receiver shown in Figure 4.7(a) performs *matched filtering*. The impulse response of the matched filter is matched to the impulse response of the flat-fading channel as h^*. The output of the matched filter is scaled down by a factor of $||h||^2$ which is the total-energy contained in the impulse response of the flat-fading channel. The resulting *decision vector* **y** serves as the *sufficient statistic* for the estimation of **s** from the received signal **r** (refer equation A.77 in [2])

$$\tilde{\mathbf{y}} = \frac{h^*}{||h||^2}\mathbf{r} = \frac{h^*}{||h||^2}h\mathbf{s} + \frac{h^*}{||h||^2}\mathbf{n} = \mathbf{s} + \tilde{\mathbf{w}} \qquad (4.12)$$

Since the absolute value $|h|$ and the Eucliden norm $||h||$ are related as $|h|^2 = ||h|| = hh^*$, the model can be simplified further as given in Figure 4.7(b). To simulate flat fading, the values for the fading variable h are

4.2 Fading channels

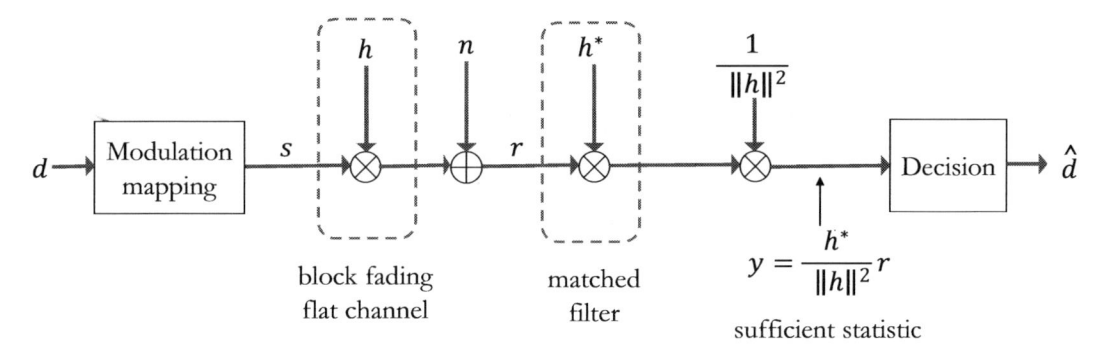

![Figure 4.7a: Estimation in additive Gaussian noise in a complex vector space showing signal flow from d through Modulation mapping, block fading flat channel (multiplied by h, added noise n), matched filter (h*), divided by $\|h\|^2$, with sufficient statistic $y = \frac{h^*}{\|h\|^2} r$, to Decision producing $\hat{d}$.]

(a) Estimation in additive Gaussian noise in a complex vector space

![Figure 4.7b: Equivalent model for flat-fading system simulation showing signal flow from d through Modulation mapping, multiplied by |h|, added noise n, multiplied by 1/|h|, to Decision producing $\hat{d}$.]

(b) Equivalent model for flat-fading system simulation

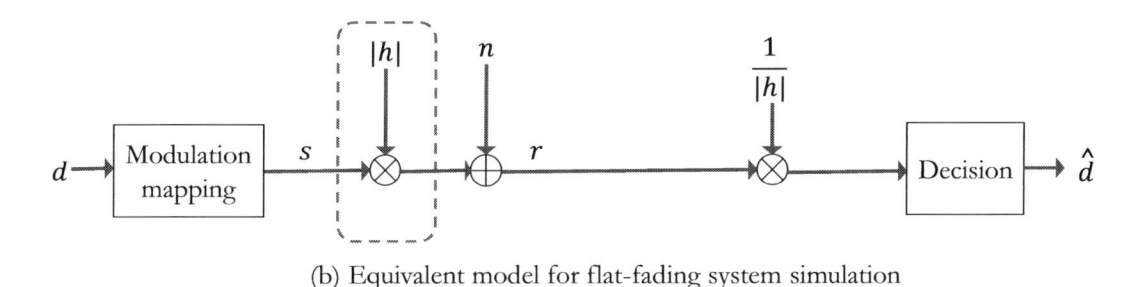

Fig. 4.7: Simulation model for modulation and detection over flat fading channel

drawn from complex normal distribution

$$h = X + jY \tag{4.13}$$

where X, Y are statistically independent real valued normal random variables.

- If $E[h] = 0$, then $|h|$ is Rayleigh distributed, resulting in a Rayleigh flat fading channel
- If $E[h] \neq 0$, then $|h|$ is Rician distributed, resulting in a Rician flat fading channel with the factor $K = [E[h]]^2/\sigma_h^2$

4.2.3 Rayleigh flat-fading channel

With respect to the simulation model shown in Figure 4.7(b), the samples for the Rayleigh flat-fading samples are drawn from the following random variable

$$h = |X + jY| \tag{4.14}$$

where $X \sim \mathcal{N}(0, \sigma^2/2)$ and $Y \sim \mathcal{N}(0, \sigma^2/2)$. The average power in the distribution is $P_{av} = \sigma^2$. Therefore to model a channel with $P_{av} = 1$, the normal random variables X and Y should have the standard deviation $\sigma = 1/\sqrt{2}$. In Python, a Rayleigh flat fading channel for N channel realizations can be simulated by the following code

Program 61: DigiCommPy\channels.py: Generating channel samples for Rayleigh flat-fading

```
def rayleighFading(N):
    """
    Generate Rayleigh flat-fading channel samples
    Parameters:
        N : number of samples to generate
    Returns:
        abs_h : Rayleigh flat fading samples
    """
    # 1 tap complex gaussian filter
    h = 1/sqrt(2)*(standard_normal(N)+1j*standard_normal(N))
    return abs(h)
```

4.2.3.1 Theoretical symbol error rates

The theoretical average probability of symbol errors over Rayleigh fading channel with AWGN noise can be obtained using *moment generating function* (MGF) approach [3][4]. Table 4.2 lists the theoretical symbol error rates for various modulations over Rayleigh fading channel with AWGN noise. The parameter $\bar{\gamma}_s$ denotes the average SNR-per-symbol (E_s/N_0) and the *moment generating function* for Rayleigh distribution is defined as

$$\mathcal{M}_{\gamma_s}\left(-\frac{g}{sin^2\phi}\right) = \left(1 + \frac{g\bar{\gamma}_s}{sin^2\phi}\right)^{-1} \tag{4.15}$$

Modulation	Parameter g	Avg Prob. of Symbol error $\bar{P}_s$
BPSK	-	$0.5\left(1 - \sqrt{\frac{\bar{\gamma}_s}{1+\bar{\gamma}_s}}\right)$
MPSK	$sin^2\left(\frac{\pi}{M}\right)$	$\frac{1}{\pi}\int_0^{\frac{(M-1)\pi}{M}} \mathcal{M}_{\gamma_s}\left(-\frac{g}{sin^2\phi}\right)d\phi$
MQAM	$\frac{1.5}{(M-1)}$	$\frac{4}{\pi}\left(1-\frac{1}{\sqrt{M}}\right)\int_0^{\pi/2} \mathcal{M}_{\gamma_s}\left(-\frac{g}{sin^2\phi}\right)d\phi - \frac{4}{\pi}\left(1-\frac{1}{\sqrt{M}}\right)^2\int_0^{\pi/4} \mathcal{M}_{\gamma_s}\left(-\frac{g}{sin^2\phi}\right)d\phi$
MPAM	$\frac{3}{(M^2-1)}$	$\frac{2(M-1)}{M\pi}\int_0^{\pi/2} \mathcal{M}_{\gamma_s}\left(-\frac{g}{sin^2\phi}\right)d\phi$

Table 4.2: Theoretical symbol error rates for various digital modulation schemes

The following Python function computes the theoretical symbol errors for various modulation schemes over a Rayleigh fading channel with AWGN noise.

Program 62: DigiCommPy\errorRates.py: Theoretical symbol error rates in Rayleigh fading channel

```
def ser_rayleigh(EbN0dBs,mod_type=None,M=0):
    """
    Theoretical Symbol Error Rates for various modulations over noise added Rayleigh
    flat-fading channel
    Parameters:
        EbN0dBs : list of SNR per bit values in dB scale
```

4.2 Fading channels

```python
            mod_type : 'PSK','QAM','PAM'
            M : Modulation level for the chosen modulation.
                For PSK,PAM M can be any power of 2.
                For QAM M must be even power of 2 (square QAM only)
        Returns:
            SERs = Symbol Error Rates
        """
        if mod_type==None:
            raise ValueError('Invalid value for mod_type')
        if (M<2) or ((M & (M -1))!=0): #if M not a power of 2
            raise ValueError('M should be a power of 2')
        func_dict = {'psk': psk_rayleigh,'qam':qam_rayleigh,'pam':pam_rayleigh}
        gamma_s_vals = log2(M)*(10**(EbN0dBs/10))
        return func_dict[mod_type.lower()](M,gamma_s_vals) #call appropriate function

def mgf_rayleigh(g,gamma_s): #MGF function for Rayleigh channel
    fun = lambda x: 1/(1+(g*gamma_s/(sin(x)**2))) # MGF function
    return fun

def psk_rayleigh(M,gamma_s_vals):
    gamma_b = gamma_s_vals/log2(M)
    if (M==2):
        SERs = 0.5*(1-sqrt(gamma_b/(1+gamma_b)))
    else:
        SERs = np.zeros(len(gamma_s_vals))
        g = (sin(pi/M))**2
        for i, gamma_s in enumerate(gamma_s_vals):
            (y,_) = quad(mgf_rayleigh(g,gamma_s),0,pi*(M-1)/M) #integration
            SERs[i] = (1/pi)*y
    return SERs

def qam_rayleigh(M,gamma_s_vals):
    if (M==1) or (np.mod(np.log2(M),2)!=0): # M not a even power of 2
        raise ValueError('Only square MQAM supported. M must be even power of 2')
    SERs = np.zeros(len(gamma_s_vals))
    g = 1.5/(M-1)
    for i, gamma_s in enumerate(gamma_s_vals):
        fun = mgf_rayleigh(g,gamma_s) # MGF function
        (y1,_) = quad(fun,0,pi/2) #integration 1
        (y2,_) = quad(fun,0,pi/4) #integration 2
        SERs[i] = 4/pi*(1-1/sqrt(M))*y1-4/pi*(1-1/sqrt(M))**2*y2
    return SERs

def pam_rayleigh(M,gamma_s_vals):
    SERs = np.zeros(len(gamma_s_vals))
    g = 3/(M**2-1)
    for i, gamma_s in enumerate(gamma_s_vals):
        (y1,_) = quad(mgf_rayleigh(g,gamma_s),0,pi/2) #integration
        SERs[i] = 2*(M-1)/(M*pi)*y1
    return SERs
```

4.2.3.2 Simulation code and performance results

In chapter 3, the code implementation for complex baseband models for various digital modulators and demodulator are given. The computation and generation of AWGN noise is given in section 4.1.2. Using these models, we can create a unified simulation code for simulating the performance of various modulation techniques over Rayleigh flat-fading channel (refer Figure 4.7(b)).

The Python code implementing the aforementioned simulation model is given next. An unified approach is employed to simulate the performance of any of the given modulation technique - MPSK, MQAM or MPAM. To simulate the performance for a given modulation, set the mod_type variable to 'PAM' or 'PSK' or 'QAM'.

The simulation code will automatically choose the selected modulation type, performs Monte Carlo simulation, computes symbol error rates and plots them against the theoretical symbol error rates. The simulated performance results obtained for various modulations are shown in the Figure 4.8.

Program 63: DigiCommPy\chapter_4\rayleighPerformance.py: Performance in Rayleigh flat fading

```python
import numpy as np #for numerical computing
import matplotlib.pyplot as plt #for plotting functions
from matplotlib import cm # colormap for color palette
from scipy.special import erfc
from DigiCommPy.modem import PSKModem,QAMModem,PAMModem,FSKModem
from DigiCommPy.channels import awgn,rayleighFading
from DigiCommPy.errorRates import ser_rayleigh

#---------Input Fields----------------------
nSym = 10**6 # Number of symbols to transmit
EbN0dBs = np.arange(start=-4,stop = 12, step = 2) # Eb/N0 range in dB for simulation
mod_type = 'PAM' # Set 'PSK' or 'QAM' or 'PAM
arrayOfM = [2,4,8,16,32] # array of M values to simulate
#arrayOfM=[4,16,64,256] # uncomment this line if MOD_TYPE='QAM'

modem_dict = {'psk': PSKModem,'qam':QAMModem,'pam':PAMModem}
colors = plt.cm.jet(np.linspace(0,1,len(arrayOfM))) # colormap
fig, ax = plt.subplots(nrows=1,ncols = 1)

for i, M in enumerate(arrayOfM):
    k=np.log2(M)
    EsN0dBs = 10*np.log10(k)+EbN0dBs # EsN0dB calculation
    SER_sim = np.zeros(len(EbN0dBs)) # simulated Symbol error rates
    # uniform random symbols from 0 to M-1
    inputSyms = np.random.randint(low=0, high = M, size=nSym)

    modem = modem_dict[mod_type.lower()](M)#choose a modem from the dictionary
    modulatedSyms = modem.modulate(inputSyms) #modulate

    for j,EsN0dB in enumerate(EsN0dBs):
        h_abs = rayleighFading(nSym) #Rayleigh flat fading samples
        hs = h_abs*modulatedSyms #fading effect on modulated symbols
        receivedSyms = awgn(hs,EsN0dB) #add awgn noise

        y = receivedSyms/h_abs # decision vector
        detectedSyms = modem.demodulate(y) #demodulate (Refer Chapter 3)
```

4.2 Fading channels

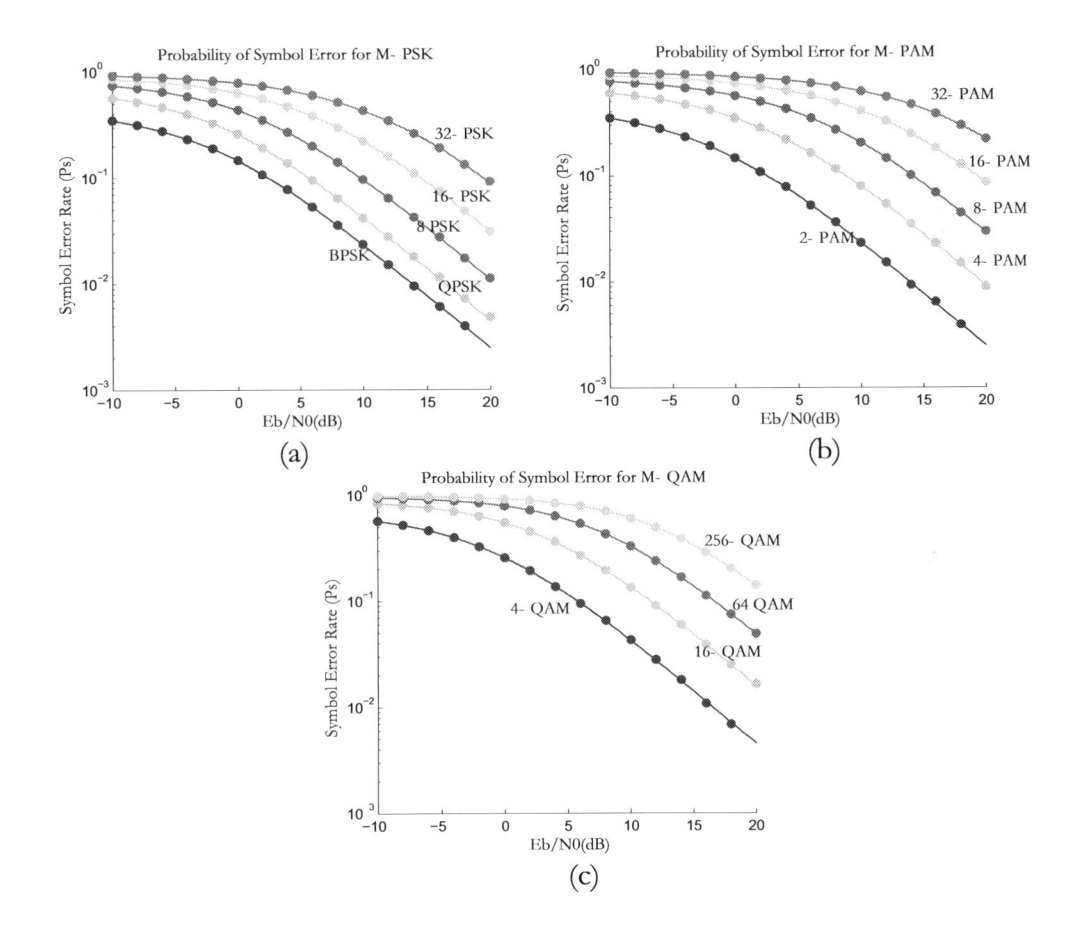

```
            SER_sim[j] = np.sum(detectedSyms != inputSyms)/nSym

        SER_theory = ser_rayleigh(EbN0dBs,mod_type,M) #theory SER
        ax.semilogy(EbN0dBs,SER_sim,color =
         ↪    colors[i],marker='o',linestyle='',label='Sim '+str(M)+'-'+mod_type.upper())
        ax.semilogy(EbN0dBs,SER_theory,color = colors[i],linestyle='-',label='Theory
         ↪    '+str(M)+'-'+mod_type.upper())

    ax.set_xlabel('Eb/N0(dB)');ax.set_ylabel('SER ($P_s$)')
    ax.set_title('Probability of Symbol Error for M-'+str(mod_type)+' over Rayleigh
     ↪    flat fading channel')
    ax.legend();fig.show()
```

Fig. 4.8: Performance of modulations over Rayleigh flat fading channel : (a) MPSK , (b) MPAM, (c) MQAM

4.2.4 Rician flat-fading channel

In wireless environments, transmitted signal may be subjected to multiple scatterings before arriving at the receiver. This gives rise to random fluctuations in the received signal and this phenomenon is called *fading*. The scattered version of the signal is designated as *non line of sight* (NLOS) component. If the number of NLOS components are sufficiently large, the fading process is approximated as the sum of large number of complex Gaussian process whose probability-density-function follows Rayleigh distribution.

Rayleigh distribution is well suited for the absence of a dominant *line of sight* (LOS) path between the transmitter and the receiver. If a line of sight path do exist, the envelope distribution is no longer Rayleigh, but Rician. If there exists a dominant LOS component, the fading process can be represented as the sum of complex exponential and a narrowband complex Gaussian process $g(t)$. If the LOS component arrive at the receiver at an *angle of arrival* (AoA) θ, phase ϕ and with the maximum Doppler frequency f_D, the fading process in baseband can be represented as [5]

$$h(t) = \underbrace{\sqrt{\frac{K\Omega}{K+1}}}_{A:=} e^{(j2\pi f_D cos(\theta)t + \phi)} + \underbrace{\sqrt{\frac{\Omega}{K+1}}}_{S:=} g(t) \qquad (4.16)$$

where, K represents the *Rician K factor* given as the ratio of power of the LOS component (A^2) to the power of the scattered components (S^2) marked in the equation above.

$$K = \frac{A^2}{S^2} \qquad (4.17)$$

The received signal power Ω is the sum of power in LOS component and the power in scattered components, given as $\Omega = A^2 + S^2$. The above mentioned fading process is called Rician fading process. The best- and worst-case Rician fading channels are associated with $K = \infty$ and $K = 0$ respectively. A Rician fading channel with $K = \infty$ is a Gaussian channel with a strong LOS path. Rician channel with $K = 0$ represents a Rayleigh channel with no LOS path.

The statistical model for generating flat-fading Ricean samples is discussed in detail in reference [1]. With respect to the simulation model shown in Figure 4.7(b), given a K factor, the samples for the Rician flat-fading samples are drawn from the following random variable

$$h = |X + jY| \qquad (4.18)$$

where $X, Y \sim \mathcal{N}(\mu, \sigma^2)$ are Gaussian random variables with non-zero mean μ and standard deviation σ as given in [6], [7]

$$\mu = \sqrt{\frac{K}{2(K+1)}} \qquad \sigma = \sqrt{\frac{1}{2(K+1)}} \qquad (4.19)$$

Program 64: DigiCommPy\channels.py: Generating channel samples for Rician flat-fading

```
def ricianFading(K_dB,N):
    """
    Generate Rician flat-fading channel samples
    Parameters:
        K_dB: Rician K factor in dB scale
        N : number of samples to generate
    Returns:
        abs_h : Rician flat fading samples
    """
```

4.2 Fading channels

```
    K = 10**(K_dB/10) # K factor in linear scale
    mu = sqrt(K/(2*(K+1))) # mean
    sigma = sqrt(1/(2*(K+1))) # sigma
    h = (sigma*standard_normal(N)+mu)+1j*(sigma*standard_normal(N)+mu)
    return abs(h)
```

4.2.4.1 Theoretical symbol error rates

The theoretical average probability of symbol errors over Rician flat-fading channel with AWGN noise can be obtained using *moment generating function* (MGF) approach [3] [4]. Table 4.2 lists the theoretical symbol error rates for various modulations over a fading channel with AWGN noise where the parameter $\bar{\gamma}_s$ denotes the average SNR-per-symbol (E_s/N_0) and the *moment generating function* (MGF) should correspond to the following equation for the case of Rician fading.

$$\mathcal{M}_{\gamma_s}\left(-\frac{g}{\sin^2\phi}\right) = \frac{(1+K)\sin^2\phi}{(1+K)\sin^2\phi + g\bar{\gamma}_s} exp\left(-\frac{Kg\bar{\gamma}_s}{(1+K)\sin^2\phi + g\bar{\gamma}_s}\right) \quad (4.20)$$

The following Python function computes the theoretical symbol errors for various modulation schemes over a Rician fading channel with AWGN noise.

Program 65: DigiCommPy\errorRates.py: Theoretical symbol error rates over Rician fading channel

```python
def ser_rician(K_dB,EbN0dBs,mod_type=None,M=0):
    """
    Theoretical Symbol Error Rates for various modulations over noise added Rician
    flat-fading channel
    Parameters:
        K_dB: Rician K-factor in dB
        EbN0dBs : list of SNR per bit values in dB scale
        mod_type : 'PSK','QAM','PAM','FSK'
        M : Modulation level for the chosen modulation.
            For PSK,PAM M can be any power of 2.
            For QAM M must be even power of 2 (square QAM only)
    Returns:
        SERs = Symbol Error Rates
    """
    if mod_type==None:
        raise ValueError('Invalid value for mod_type')
    if (M<2) or ((M & (M -1))!=0): #if M not a power of 2
        raise ValueError('M should be a power of 2')
    
    func_dict = {'psk': psk_rician,'qam':qam_rician,'pam':pam_rician}
    gamma_s_vals = log2(M)*(10**(EbN0dBs/10))
    #call appropriate function
    return func_dict[mod_type.lower()](K_dB,M,gamma_s_vals)

def mgf_rician(K_dB,g,gamma_s): #MGF function for Rician channel
    K = 10**(K_dB/10) # K factor in linear scale
    fun = lambda x: ((1+K)*sin(x)**2)/((1+K)*sin(x)**2+g*gamma_s)\
```

```python
            *exp(-K*g*gamma_s/((1+K)*sin(x)**2+g*gamma_s)) # MGF function
    return fun #return the MGF function

def psk_rician(K_dB,M,gamma_s_vals):
    gamma_b = gamma_s_vals/log2(M)

    if (M==2):
        SERs = 0.5*(1-sqrt(gamma_b/(1+gamma_b)))
    else:
        SERs = np.zeros(len(gamma_s_vals))
        g = (sin(pi/M))**2
        for i, gamma_s in enumerate(gamma_s_vals):
            (y,_) = quad(mgf_rician(K_dB,g,gamma_s),0,pi*(M-1)/M) #integration
            SERs[i] = (1/pi)*y
    return SERs

def qam_rician(K_dB,M,gamma_s_vals):
    if (M==1) or (np.mod(np.log2(M),2)!=0): # M not a even power of 2
        raise ValueError('Only square MQAM supported. M must be even power of 2')
    SERs = np.zeros(len(gamma_s_vals))
    g = 1.5/(M-1)
    for i, gamma_s in enumerate(gamma_s_vals):
        fun = mgf_rician(K_dB,g,gamma_s) #MGF function
        (y1,_) = quad(fun,0,pi/2) #integration 1
        (y2,_) = quad(fun,0,pi/4) #integration 2
        SERs[i] = 4/pi*(1-1/sqrt(M))*y1-4/pi*(1-1/sqrt(M))**2*y2
    return SERs

def pam_rician(K_dB,M,gamma_s_vals):
    SERs = np.zeros(len(gamma_s_vals))
    g = 3/(M**2-1)
    for i, gamma_s in enumerate(gamma_s_vals):
        (y1,_) = quad(mgf_rician(K_dB,g,gamma_s),0,pi/2) #integration
        SERs[i] = 2*(M-1)/(M*pi)*y1
    return SERs
```

4.2.4.2 Simulation code and performance results

In chapter 3, the code implementation for complex baseband models for various digital modulators and demodulator are given. The computation and generation of AWGN noise is given section 4.1.2. Use these models, we can create a unified simulation for code for simulating the performance of various modulation techniques over Rician flat-fading channel the simulation model shown in Figure 4.7(b).

The Python code implementing the aforementioned simulation model is given next. An unified approach is employed to simulate the performance of any of the given modulation technique - MPSK, MQAM or MPAM. To simulate the performance for a given modulation, the user just needs to set the mod_type variable to 'PAM' or 'PSK' or 'QAM'.

4.2 Fading channels

The simulation code will automatically choose the selected modulation type, performs Monte Carlo simulation, computes symbol error rates and plots them against the theoretical symbol error rate curves. The simulated performance results obtained for various modulations are shown in the Figure 4.9.

Program 66: DigiCommPy\chapter_4\ricianPerformance.py: Performance over Rician flat fading

```python
import numpy as np #for numerical computing
import matplotlib.pyplot as plt #for plotting functions
from matplotlib import cm # colormap for color palette
from scipy.special import erfc
from DigiCommPy.modem import PSKModem,QAMModem,PAMModem,FSKModem
from DigiCommPy.channels import awgn,ricianFading
from DigiCommPy.errorRates import ser_rician

#---------Input Fields----------------------
nSym = 10**6 # Number of symbols to transmit
EbN0dBs = np.arange(start=0,stop = 22, step = 2) # Eb/N0 range in dB for simulation
K_dBs = [3,5,10,20] # array of K factors for Rician fading in dB
mod_type = 'PSK' # Set 'PSK' or 'QAM' or 'PAM'
M = 4 # M value for the modulation to simulate

modem_dict = {'psk': PSKModem,'qam':QAMModem,'pam':PAMModem}
colors = plt.cm.jet(np.linspace(0,1,len(K_dBs))) # colormap
fig, ax = plt.subplots(nrows=1,ncols = 1)

for i, K_dB in enumerate(K_dBs):
    #-----Initialization of various parameters----
    k=np.log2(M)
    EsN0dBs = 10*np.log10(k)+EbN0dBs # EsN0dB calculation
    SER_sim = np.zeros(len(EbN0dBs)) # simulated Symbol error rates
    # uniform random symbols from 0 to M-1
    inputSyms = np.random.randint(low=0, high = M, size=nSym)
    
    modem = modem_dict[mod_type.lower()](M)#choose a modem from the dictionary
    modulatedSyms = modem.modulate(inputSyms) #modulate
    
    for j,EsN0dB in enumerate(EsN0dBs):
        h_abs = ricianFading(K_dB,nSym) #Rician flat fading samples
        hs = h_abs*modulatedSyms #fading effect on modulated symbols
        receivedSyms = awgn(hs,EsN0dB) #add awgn noise
        y = receivedSyms/h_abs # decision vector
        detectedSyms = modem.demodulate(y) #demodulate (Refer Chapter 3)
        SER_sim[j] = np.sum(detectedSyms != inputSyms)/nSym
    
    SER_theory = ser_rician(K_dB,EbN0dBs,mod_type,M)
    ax.semilogy(EbN0dBs,SER_sim,color =
     ↪ colors[i],marker='o',linestyle='',label='Sim K='+str(K_dB)+' dB')
    ax.semilogy(EbN0dBs,SER_theory,color = colors[i],linestyle='-',label='Theory
     ↪ K='+str(K_dB)+' dB')
```

```
ax.set_xlabel('Eb/N0(dB)');ax.set_ylabel('SER ($P_s$)')
ax.set_title('Probability of Symbol Error for M-'+str(mod_type)+' over Rayleigh
↪ flat fading channel');ax.legend();fig.show()
```

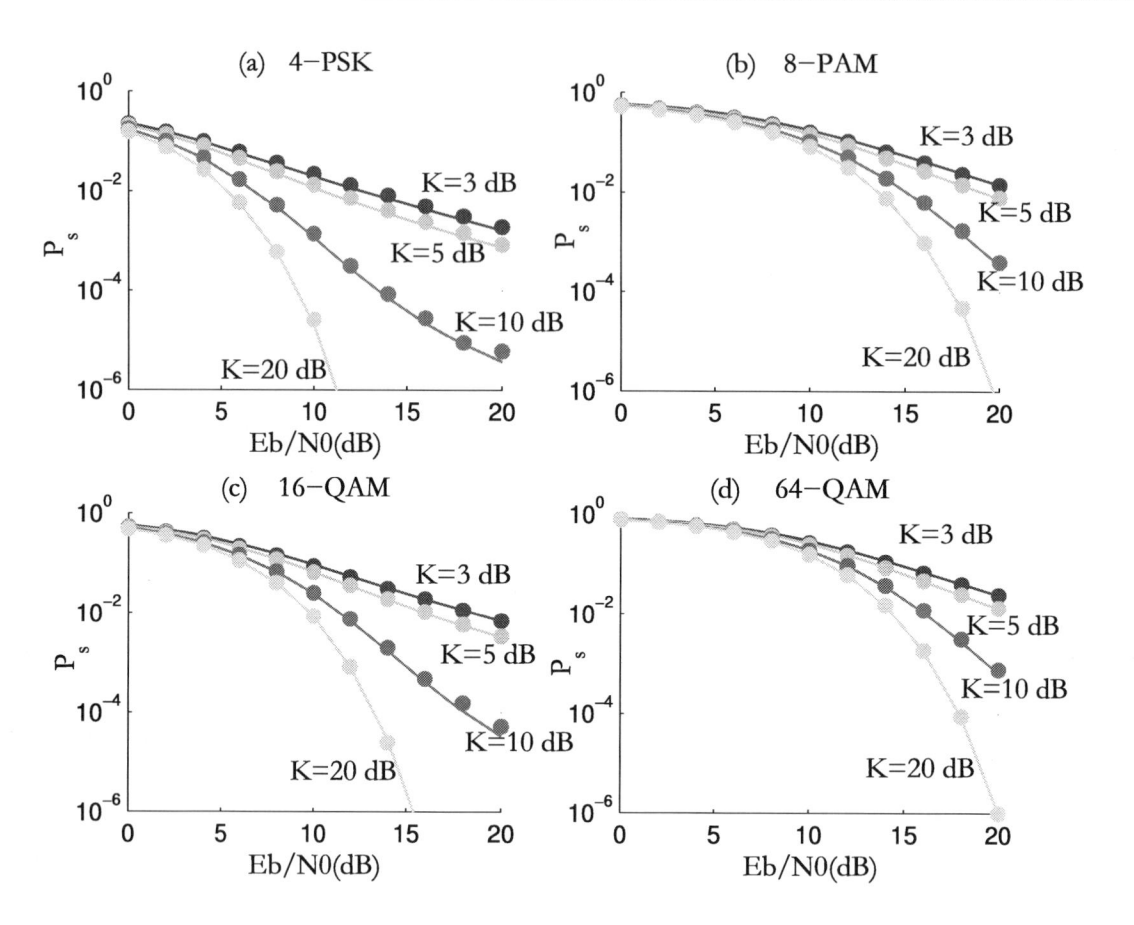

Fig. 4.9: Performance of modulations over Rician flat fading channel with different K factors : (a) QPSK , (b) 8-PAM, (c) 16-QAM, (d) 64-QAM

References

1. Mathuranathan Viswanathan, *Wireless Communication Systems in Matlab*, ISBN 978-1720114352, September 2018.
2. D. Tse and P. Viswanath, *Fundamentals of Wireless Communication*, Cambridge University Press, 2005
3. Andrea Goldsmith, *Wireless Communications*, Cambridge University Pres, first edition, August 8, 2005
4. M. K. Simon and M.-S. Alouini, *Digital Communication over Fading Channels A Unified Approach to Performance Analysis*, Wiley 2000
5. C. Tepedelenlioglu, A. Abdi, and G. B. Giannakis, *The Ricean K factor: Estimation and performance analysis*, IEEE Trans. Wireless Communication ,vol. 2, no. 4, pp. 799–810, Jul. 2003.
6. R. F. Lopes, I. Glover, M. P. Sousa, W. T. A. Lopes, and M. S. de Alencar, *A software simulation framework for spectrum sensing*, 13th International Symposium on Wireless Personal Multimedia Communications (WPMC 2010), Out. 2010.
7. M. C. Jeruchim, P. Balaban, and K. S. Shanmugan, *Simulation of Communication Systems, Methodology, Modeling, and Techniques*, second edition Kluwer Academic Publishers, 2000.

Chapter 5
Linear Equalizers

Abstract In communication systems that are dominated by inter-symbol interference, equalizers play an important role in combating unwanted distortion effects induced by the channel. Linear equalizers are based on linear time-invariant filters that are simple to analyze using conventional signal processing theory. This chapter is focused on design and implementation of three important types of linear equalizers namely, the *zero-forcing* equalizer, the *minimum mean squared error* equalizer and the *least mean square* adaptive equalizer.

5.1 Introduction

Transmission of signals through band-limited channels are vulnerable to time dispersive effects of channel that could manifest as *intersymbol interference* (ISI). Three major approaches to deal with ISI are:

- *Nyquist first criterion*: Force ISI effect to zero by signal design - pulse shaping techniques like sinc filtering, raised-cosine filtering and square-root raised-cosine filtering.
- *Partial response signaling*: Introduce controlled amount of ISI in the transmit side and prepare to deal with it at the receiver.
- *Design algorithms to counter ISI*: Learn to live with the presence of ISI and design robust algorithms at the receiver - Viterbi algorithm (maximum likelihood sequence estimation), equalizer etc.,

This chapter focuses on the third item, particularly on the technique of employing an *equalizer* at the receiver. An equalizer is a digital filter at the receiver that mitigates the distortion effects of *intersymbol interference* (ISI), introduced by a time-dispersive channel. If the channel is unknown and time varying, optimum design of transmit and receive filters is not possible. However, one can neutralize (equalize) the channel effects, if the impulse response of the channel is made known at the receiver. The problem of recovering the input signal in a time-dispersive channel boils down to finding the inverse of the channel and using it to neutralize the channel effects. However, finding the inverse of the channel impulse response may be difficult due to the following reasons:

- If the channel response is zero at any frequency, the inverse of the channel response is undefined at that frequency.
- Generally, the receiver does not know the channel response and extra circuitry (channel estimation) may be needed to estimate it.
- The channel can be varying in real time and therefore, the equalization needs to be adaptive.

Additionally, even if the channel response is of finite length, the equalizer can have infinite impulse response. Usually, from stability perspective, the equalizer response has to be truncated to a finite length. As a result, perfect equalization is difficult to achieve and does not always provide the best performance.

Classification of equalizer structures

Based on implementation structures, the equalizer structures are categorized as follows:
- Maximum likelihood sequence estimation (MLSE) equalizer
  - Provides the most optimal performance.
  - Computational complexity can be extremely high for practical applications.
- Decision feedback equalizer (DFE)
  - A nonlinear equalizer that employs previous decisions as training sequences.
  - Provides better performance but suffers from the serious drawback of error propagation in the event of erroneous decisions.
  - To provide robustness, it is often combined with training sequence based feed-forward equalization.
- Linear equalizer
  - The most suboptimum equalizer with lower complexity.

5.2 Linear equalizers

The MLSE equalizers provide the most optimum performance among all the equalizer categories. For long channel responses, the MLSE equalizers become too complex and hence, from practical implementation point of view, other suboptimum filters are preferred. The most suboptimum equalizer is the *linear equalizer* that offers lower complexity. Linear equalizers are particularly effective for those channels where the ISI is not severe. Commonly, equalizers are implemented as a digital filter structure having a number of taps with complex tap coefficients (also referred as tap weights).

Choice of adaptation of equalizer tap weights

Based on the adaptation of tap coefficients, the linear equalizers can be classified into two types.

- *Preset equalizers*: If the channel is considered to be time-invariant during the data transmission interval, the equalizer weights are calculated during the beginning of the transmission session and are fixed during the rest of the transmission session.
- *Adaptive equalizers*: For time variant channels, the coefficients are computed using an adaptive algorithm to adapt for the change in the channel conditions, thus rendering the name. There exist numerous choices of algorithms for calculating and updating the adaptive weights - *least mean squares*, *recursive lattice square*, *conventional Kalman*, *square-root Kalman* and *fast Kalman* - to name a few.

Choice of sampling time

The choice of sampling time with respect to the symbol time period, affects the design and performance of equalizers. As a consequence, two types of equalization techniques exist.

- *Symbol-spaced linear equalizer*: Shown in Figure 5.1, where, the downsampler in front of the equalizer, operates at 1-sample per symbol time (T_{sym}). These equalizers are very sensitive to sampling time, potentially introducing destructive aliasing effects and therefore the performance may not be optimal.

5.2 Linear equalizers

- *Fractionally-spaced linear equalizer*: Here, the idea is to reduce the space between the adjacent equalizer taps to a fraction of the symbol interval. Equivalently, the downsampler that precedes the equalizer operates at a rate higher than symbol rate (say spacing in time kT_{sym}/M - for some integers k and M). Finally, the equalizer output is downsampled by symbol time, before delivering the output symbols to the detector. This is illustrated in Figure 5.2. Though it requires more computation, fractionally-spaced equalizers simplify the rest of the demodulator design [1] and are able to compensate for any arbitrary sampling time phase [2].

This chapter deals only with symbol-spaced linear equalizer with preset tap weights and adaptive weights using least mean squares algorithm. The treatment of fractionally-spaced equalizers is beyond the scope of this book.

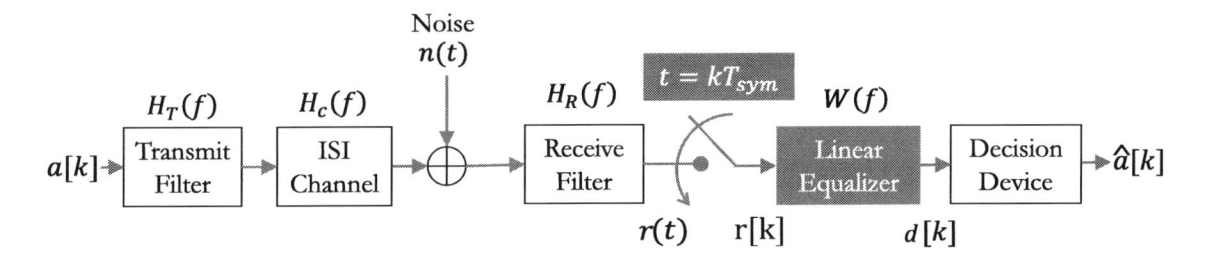

Fig. 5.1: Continuous-time channel model with symbol-spaced linear equalizer

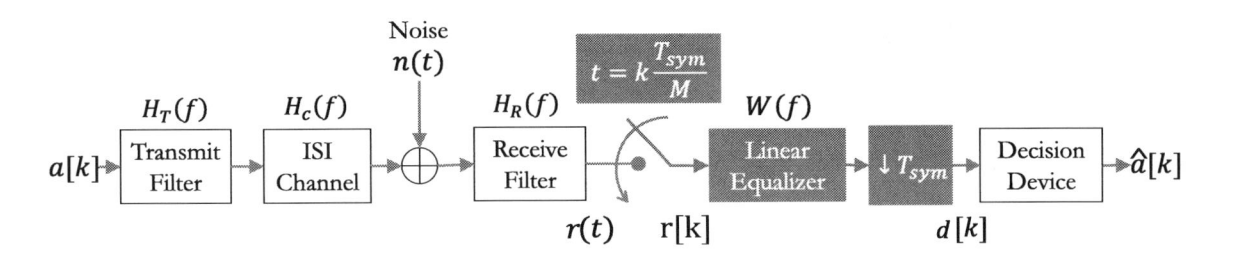

Fig. 5.2: Continuous-time channel model with fractionally-spaced linear equalizer

Choice of criterion for tap weights optimization

The equalizer tap weights are usually chosen based on some optimal criterion of which two of them are given.

- *Peak distortion criterion*: Aims at minimizing the maximum possible distortion of a signal, caused by ISI, at the output of the equalizer. This is also equivalent to *zero-forcing criterion*. This forms the basis for *zero-forcing equalizers* [3].
- *Mean square error criterion*: Attempts to minimize the mean square value of an error term computed from the difference between the equalizer output and the transmitted information symbol. This forms the basis for *linear minimum mean square error (LMMSE)* equalizer and *least mean square (LMS)* algorithm.

5.3 Symbol-spaced linear equalizer channel model

Consider the continuous-time channel model in Figure 5.1. In this generic representation, the transmit pulse shaping filter and the receiver filter are represented by the filters $H_T(f)$ and $H_R(f)$ respectively. The transmit filter may or may not be a square-root raised cosine filter. However, we assume that the receive filter is a square-root raised cosine filter. Therefore, the sampled noise $n[k] = h_R(t) * n(t) |_{t=kT_{sym}}$ is white Gaussian.

In general, the magnitude response of the channel $|H_c(f)|$ is not a constant over a range of frequencies channel, that is the impulse response of the channel $h_c(t)$ is not ideal. Therefore, linear distortions become unavoidable.

We designate the combined effect of transmit filter, channel and the receive filter as overall channel. The overall impulse response $h(t)$ is given by

$$h(t) = h_T(t) * h_c(t) * h_R(t) \tag{5.1}$$

The received signal after the symbol rate sampler is given by

$$r[k] = \sum_{i=-\infty}^{\infty} h[i]a[k-i] + z[k] \tag{5.2}$$

The resulting simplified discrete-time channel model is given in Figure 5.3, where $a[k]$ represent the information symbols transmitted through a channel having an arbitrary impulse response $h[k]$, $r[k]$ are the received symbols at the receiver and $n[k]$ is white Gaussian noise.

$$r[k] = \sum_{i=-\infty}^{\infty} h[i]a[k-i] + n[k] \tag{5.3}$$

Usually, in practice, the channel impulse response can be truncated to some finite length L. Under assumptions of causality of transmit filter, channel and the receive filter, we can safely hold that $h[i] = 0$ for $i < 0$. If L is sufficiently large, $h[i] \approx 0$ holds for $i \geq L$. Hence the received signal can be rewritten as

$$r[k] = \sum_{i=0}^{L-1} h[i]a[k-i] + n[k] \tag{5.4}$$

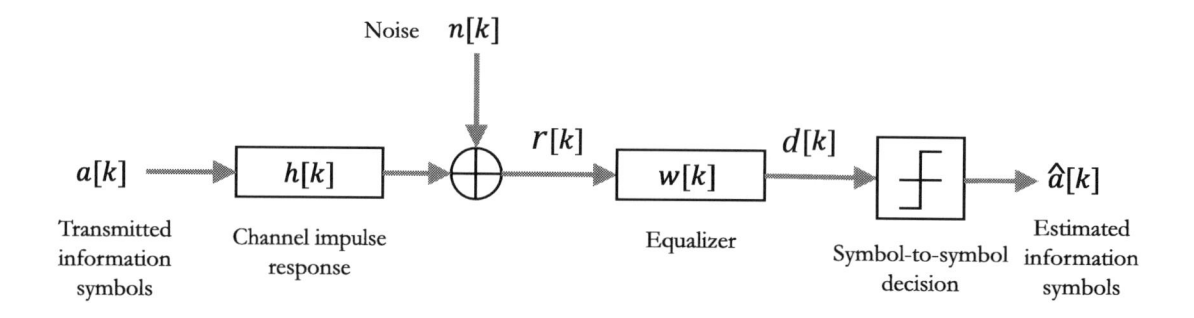

Fig. 5.3: Discrete-time channel model for symbol-spaced linear equalization

The equalizer co-efficients are represented by $w[k]$. Symbol-by-symbol decisions are made on the equalizer output $d[k]$ and the transmitted information symbols are estimated as $\hat{a}[k]$.

5.4 Implementing equalizers using object oriented programming

Figure 5.4 shows the equivalent frequency domain representation of the channel model, where the frequency response can be computed using Z-transform as

$$W(z) = \sum_{k=-\infty}^{\infty} w[k] z^{-k} \qquad (5.5)$$

A linear equalizer is the most simplest type of equalizer, that attempts to invert the channel transfer function $H(z)$ and filters the received symbols with the inverted response. Subsequently, using a symbol-by-symbol decision device, the filtered equalizer outputs are used to estimate the information symbols.

The structure of a linear equalizer can be of FIR or IIR type. The equalizer tap weights are calculated based an optimization criterion. We will be focusing on *zero-forcing (ZF)* criterion and the *minimum mean square error (MMSE)* criterion.

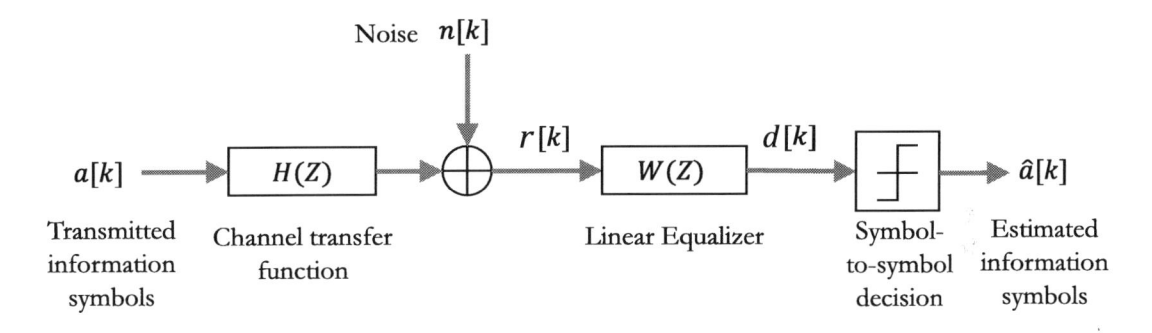

Fig. 5.4: Equivalent channel model in frequency domain with linear equalizer

5.4 Implementing equalizers using object oriented programming

In the following sections, the implementations of zero-forcing (ZF) equalizer, minimum mean square error (MMSE) equalizer and least mean square (LMS) adaptive equalizer, are described. The linear equalizers share a basic construct that can be exploited to write simplified code using *object oriented programming* in Python. We will begin by creating a common *base class* named `Equalizer` and implement the aforementioned equalization techniques as *derived classes*.

When an equalizer object is instantiated, the *constructor* defined in the base class, acts as an initializer that sets the equalizer length to the given size, initializes the equalizer weights to all zeros and sets the optimal equalizer delay to zero. The actual weights of the equalizers are determined by the type of equalizer that is designed.

The `design` member function is an abstract method whose definitions are provided by the `zeroForcing`, `MMSEEQ` and `LMSEQ` derived classes.

The `equalize` member function implements the common functionality of a linear equalizer, the process of equalization - which is, taking in an input sequence, convolving it with the equalizer tap weights and producing an equalized output. The `convMatrix` method implements the process of convolution as previously described in chapter 1 section 1.7.2.

Now that the `Equalizer` base class is defined, computing the equalize tap weights for a given equalizer design (zero-forcing, MMSE or LMS) is all that is remaining to be implemented.

Program 67: DigiCommPy\equalizers.py: Defining the Equalizer base class

```python
import numpy as np
import abc

class Equalizer():
    # Base class: Equalizer (Abstract base class)
    # Attribute definitions:
    #     self.N: length of the equalizer
    #     self.w : equalizer weights
    #     self.delay : optimized equalizer delay
    def __init__(self,N): # constructor for N tap FIR equalizer
        self.N = N
        self.w = np.zeros(N)
        self.opt_delay = 0

    @abc.abstractmethod
    def design(self): #Abstract method
        "Design the equalizer for the given impulse response and SNR"

    def convMatrix(self,h,p):
        """
        Construct the convolution matrix of size (N+p-1)x p from the
        input matrix h of size N. (see chapter 1)
        Parameters:
            h : numpy vector of length L
            p : scalar value
        Returns:
            H : convolution matrix of size (L+p-1)xp
        """
        col=np.hstack((h,np.zeros(p-1)))
        row=np.hstack((h[0],np.zeros(p-1)))

        from scipy.linalg import toeplitz
        H=toeplitz(col,row)
        return H

    def equalize(self,inputSamples):
        """
        Equalize the given input samples and produces the output
        Parameters:
            inputSamples : signal to be equalized
        Returns:
            equalizedSamples: equalized output samples
        """
        #convolve input with equalizer tap weights
        equalizedSamples = np.convolve(inputSamples,self.w)
        return equalizedSamples
```

5.5 Zero-forcing equalizer

A zero-forcing (ZF) equalizer is so named because it forces the residual ISI in the equalizer output $d[k]$ to zero. An optimum zero-forcing equalizer achieves perfect equalization by forcing the residual ISI at sampling instants kT except $k = 0$. This goal can be achieved if we allow for equalizer with infinite impulse response (IIR) structure. In most practical applications, the channel transfer function $H(z)$ can be approximated to finite response (FIR) filter and hence its perfect equalizing counterpart $W(z)$ will be an IIR filter.

$$W(z) = \frac{1}{H(z)} \tag{5.6}$$

As a result, the resulting overall transfer function, denoted by $Q(z)$, is given by

$$Q(z) = H(z)W(z) = 1 \tag{5.7}$$

Otherwise, in time-domain, the zero-forcing solution forces the overall response to zero at all positions except for the position at k_0 where its value is equal to Dirac delta function.

$$q[k] = h[k] * w[k] = \delta[k - k_0] = \begin{cases} 1, & for\ k = k_0 \\ 0, & for\ k \neq k_0 \end{cases} \tag{5.8}$$

For practical implementation, due to their inherent stability and better immunization against finite length word effects, FIR filters are practically preferred over IIR filters (refer chapter chapter 1 section 1.9 for more details). Implementing a zero-forcing equalizer as FIR filter would mean imposing causality and a length constraint on the equalizer filter whose transfer function takes the form

$$W(z) = \sum_{k=0}^{N-1} w[k] z^{-k} \tag{5.9}$$

Given the channel impulse response of length L and the equalizer filter of length N, the zero-forcing solution in equation 5.8, can be expressed as follows (refer chapter 1 section 1.6).

$$q[k] = h[k] * w[k] = \sum_{i=-\infty}^{\infty} h[i] w[k-i] = \delta[k - k_0], \quad k = 0, \cdots, L + N - 2 \tag{5.10}$$

In simple terms, with the zero-forcing solution, the overall impulse response takes the following form.

$$q[k] = h[k] * w[k] = \delta[k - k_0] = [0, 0, \cdots, 0, 1, 0, \cdots, 0, 0] \tag{5.11}$$

where, $q[k] = 1$ at position k_0.

The convolution sum of finite length sequence **h** of length L and a causal finite length sequence **w** of length N can be expressed as in matrix form as

$$\begin{bmatrix} q[0] \\ q[1] \\ q[2] \\ \vdots \\ q[L+N-2] \end{bmatrix} = \begin{bmatrix} h[0] & h[-1] & \cdots & h[-(N-1)] \\ h[1] & h[0] & \cdots & h[-(N-2)] \\ h[2] & h[1] & \cdots & h[-(N-3)] \\ \vdots & \vdots & \ddots & \vdots \\ h[L+N-2] & h[L+N-3] & \cdots & h[L-1] \end{bmatrix} \cdot \begin{bmatrix} w[0] \\ w[1] \\ w[2] \\ \vdots \\ w[N-1] \end{bmatrix} \tag{5.12}$$

Obviously, the zero-forcing solution in equation 5.8 can be expressed in matrix form as

$$\begin{bmatrix} h[0] & h[-1] & \cdots & h[-(N-1)] \\ h[1] & h[0] & \cdots & h[-(N-2)] \\ h[2] & h[1] & \cdots & h[-(N-3)] \\ \vdots & \vdots & \ddots & \vdots \\ h[L+N-2] & h[L+N-3] & \cdots & h[L-1] \end{bmatrix} \cdot \begin{bmatrix} w[0] \\ w[1] \\ w[2] \\ \vdots \\ w[N-1] \end{bmatrix} = \begin{bmatrix} 0 \\ \vdots \\ 0 \\ 1 \\ 0 \\ \vdots \\ 0 \end{bmatrix}$$

$$\mathbf{H}.\mathbf{w} = \delta_{k_0} \qquad (5.13)$$

If we assume causality of the channel impulse response, $h[k] = 0$ for $k < 0$ and if L is chosen sufficiently large $h[k] \approx 0$ holds for $k \geq L$. Then, the zero-forcing solution can be simplified as

$$\begin{bmatrix} h[0] & 0 & \cdots & 0 & 0 \\ h[1] & h[0] & \cdots & 0 & 0 \\ h[2] & h[1] & \cdots & 0 & 0 \\ \vdots & \vdots & \ddots & \vdots & \vdots \\ 0 & 0 & \ddots & h[L-1] & 0 \\ 0 & 0 & \cdots & 0 & h[L-1] \end{bmatrix} \cdot \begin{bmatrix} w[0] \\ w[1] \\ w[2] \\ \vdots \\ w[N-2] \\ w[N-1] \end{bmatrix} = \begin{bmatrix} 0 \\ \vdots \\ 0 \\ 1 \\ 0 \\ \vdots \\ 0 \end{bmatrix}$$

$$\mathbf{T}(h).\mathbf{w} = \delta_{k_0}$$
$$\mathbf{H}.\mathbf{w} = \delta_{k_0} \qquad (5.14)$$

where, $\mathbf{H} = \mathbf{T}(h)$ takes the form of a *Toeplitz matrix* (see chapter 1 section 1.7.2).

If the zero-forcing equalizer is assumed to perfectly compensate the channel impulse response, the output of the equalizer will be a Dirac delta function that is delayed by certain symbols due to equalizer delay (k_0). In matrix form, this condition is represented as

$$\mathbf{H} \cdot \mathbf{w} = \delta_{k_0} \qquad (5.15)$$

where $\mathbf{H}$ is a rectangular *channel matrix* as defined in equations 5.13 and 5.14, $\mathbf{w}$ is a column vector containing the equalizer tap weights and δ_{k_0} is a column vector representing the Dirac delta function, with all elements equal to zero except for the unity element at $(k_0)^{th}$ position. The position of the unity element in the δ_{k_0} matrix determines the equalizer delay (k_0). The equalizer delay allows for end-to-end system delay. Without the equalizer delay, it would be difficult to compensate the effective channel delay using a causal FIR equalizer.

5.5.1 Least squares solution

The method of *least squares* can be applied to solve an overdetermined system of linear equations of form $\mathbf{Hw} = \delta_{k_0}$, that is, a system in which the $\mathbf{H}$ matrix is a rectangular $(L+N-1) \times N$ matrix with more equations than unknowns $((L+N-1) > N)$.

In general, for the overdetermined $(L+N-1) \times N$ system $\mathbf{Hw} = \delta_{k_0}$, there are solutions $\mathbf{w}$ minimizing the *Euclidean distance* $\|\mathbf{Hw} - \delta_{k_0}\|^2$ and those solutions are given by the square $N \times N$ system

$$\mathbf{H}^H \mathbf{Hw} = \mathbf{H}^H \delta_{k_0} \qquad (5.16)$$

where $\mathbf{H}^H$ represents the *Hermitian transpose* (conjugate transpose of a matrix) of the channel matrix $\mathbf{H}$.

5.5 Zero-forcing equalizer

Besides, when the columns of $\mathbf{H}$ are linearly independent, it is apparent that $\mathbf{H}^H\mathbf{H}$ is symmetric and invertible, and so the solution for the zero-forcing equalizer coefficients $\mathbf{w}$ is unique and given by

$$\mathbf{w} = \left(\mathbf{H}^H\mathbf{H}\right)^{-1}\mathbf{H}^H \delta_{k_0} = \mathbf{H}^\dagger \delta_{k_0} \tag{5.17}$$

The expression $\mathbf{H}^\dagger = \left(\mathbf{H}^H\mathbf{H}\right)^{-1}\mathbf{H}^H$ is called *pseudo-inverse matrix* or *Moore-Penrose generalized matrix inverse*. To solve for equalizer taps using equation 5.17, the channel matrix $\mathbf{H}$ has to be estimated.

The *mean square error (MSE)* between the equalized samples and the ideal samples is calculated as [5]

$$\xi_N = 1 - \sum_{i=0}^{k_0} w_N(i) h_c^*(k_0 - i) \tag{5.18}$$

The equations for zero-forcing equalizer have the same form as that of MMSE equalizer (see equations 5.39 and 5.40), except for the noise term. Therefore, the MSE of the ZF equalizer can be written as [6] [7]

$$\xi_{min} = \sigma_a^2 \left(1 - \delta_{k_0}^T \mathbf{H}\mathbf{H}^\dagger \delta_{k_0}\right) \tag{5.19}$$

The optimal-delay that minimizes the MSE is simply the index of the maximum diagonal element of the matrix $\mathbf{H}\mathbf{H}^\dagger$

$$optimum\ delay = k_{opt} = \arg\max_{index}\left[diag\left(\mathbf{H}\mathbf{H}^\dagger\right)\right] \tag{5.20}$$

5.5.2 Noise enhancement

From the given discrete-time channel model (Figure 5.3), the received sequence $\mathbf{r}$ can be computed as the convolution sum of the input sequence $\mathbf{a}$ of length N and the channel impulse response $\mathbf{h}$ of length L

$$r[k] = h[k] * a[k] + n[k] = \sum_{i=-\infty}^{\infty} h[i] a[k-i] + n[k] \qquad k = 0, 1, \cdots, L+N-2 \tag{5.21}$$

Following the derivations similar to equations 5.10 and 5.12, the convolution sum can be expressed in matrix form as

$$\begin{bmatrix} r[0] \\ r[1] \\ r[2] \\ \vdots \\ r[L+N-2] \end{bmatrix} = \begin{bmatrix} h[0] & h[-1] & \cdots & h[-(N-1)] \\ h[1] & h[0] & \cdots & h[-(N-2)] \\ h[2] & h[1] & \cdots & h[-(N-3)] \\ \vdots & \vdots & \ddots & \vdots \\ h[L+N-2] & h[L+N-3] & \cdots & h[L-1] \end{bmatrix} \cdot \begin{bmatrix} a[0] \\ a[1] \\ a[2] \\ \vdots \\ a[N-1] \end{bmatrix} + \begin{bmatrix} n[0] \\ n[1] \\ n[2] \\ \vdots \\ n[N-1] \end{bmatrix} \tag{5.22}$$

$$\mathbf{r} = \mathbf{H}.\mathbf{a} + \mathbf{n} \tag{5.23}$$

If a linear zero-forcing equalizer is used, obtaining the equalizer output is equivalent to applying the pseudo-inverse matrix $\mathbf{H}^\dagger$ on the vector of received signal sequence $\mathbf{r}$.

$$\mathbf{H}^\dagger \mathbf{r} = \mathbf{H}^\dagger \mathbf{H}.\mathbf{a} + \mathbf{H}^\dagger \mathbf{n} = \mathbf{a} + \tilde{\mathbf{n}} \tag{5.24}$$

This renders the additive white Gaussian noise become colored and thereby complicates optimal detection.

5.5.3 Design and simulation of zero-forcing equalizer

The simulation model for ZF equalizer is shown in Figure 5.3. The simulation model involves the overall continuous time channel impulse response $h(t)$ with discrete-time equivalent $h[k]$ and a discrete-time zero-forcing symbol-spaced equalizer $w[k]$. The goal is to design the zero-forcing equalizer of length N that could compensate for the distortion effects introduced by the channel of length L.

As an example for simulation, the sampling frequency of the overall system is assumed as $F_s = 100Hz$ with 5 samples per symbol ($nSamp = 5$) and a channel model with the following overall channel impulse response at baud-rate is also assumed.

$$h(t) = \frac{1}{1 + \left(\frac{t}{T_{sym}}\right)^2} \tag{5.25}$$

```
import numpy as np #for numerical computing
import matplotlib.pyplot as plt #for plotting functions
from numpy import pi,log,convolve
nSamp=5 #%Number of samples per symbol determines baud rate Tsym
Fs=100 # Sampling Frequency of the system
Ts=1/Fs # Sampling time
Tsym=nSamp*Ts # symbol time period

#Define transfer function of the channel
k=6 # define limits for computing channel response
N0 = 0.001 # Standard deviation of AWGN channel noise
t = np.arange(start=-k*Tsym,stop=k*Tsym,step=Ts) # time base defined till +/-kTsym
h_t = 1/(1+(t/Tsym)**2) # channel model, replace with your own model
h_t = h_t + N0*np.random.randn(len(h_t)) # add Noise to the channel response
h_k = h_t[0::nSamp] # downsampling to represent symbol rate sampler
t_inst=t[0::nSamp] # symbol sampling instants
```

Next, plot the channel impulse response of the above model, the resulting plot is given in Figure 5.5.

```
fig, ax = plt.subplots(nrows=1,ncols = 1)
ax.plot(t,h_t,label='continuous-time model');#response at sampling instants
# channel response at symbol sampling instants
ax.stem(t_inst,h_k,'r',label='discrete-time model',use_line_collection=True)
ax.legend();ax.set_title('Channel impulse response');
ax.set_xlabel('Time (s)');ax.set_ylabel('Amplitude');fig.show()
```

Since the channel response, plotted in Figure 5.5, is of length $L = 13$, a zero-forcing forcing filter of length $N > L$ is desired. Let's fix the length of the zero-forcing filter to $N = 14$. The zero-forcing equalizer design, follows equation 5.17 and the equalizer delay determines the position of the solitary ' 1 ' in the δ_{k_0} matrix that represents the Dirac-delta function.

For implementation, the zero-forcing filter is implemented as a class *derived* from the Equalizer class that was defined in section 5.4. The design method implements the least squares solutions (equations 5.17, 5.19 and 5.20) for computing equalizer tap weights, means squared error and the optimized equalizer delay.

5.5 Zero-forcing equalizer

Channel impulse response

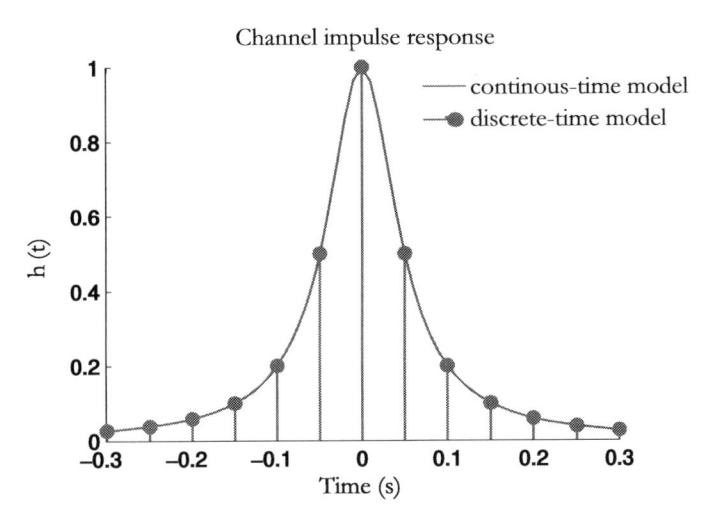

Fig. 5.5: Channel impulse response

Program 68: DigiCommPy\equalizers.py: Defining the zeroForcing equalizer derived class

```python
class zeroForcing(Equalizer): #Class zero-forcing equalizer
    def design(self,h,delay=None): #override method in Equalizer abstract class
        """
        Design a zero forcing equalizer for given channel impulse response (CIR).
        If the tap delay is not given, a delay optimized equalizer is designed
        Parameters:
            h : channel impulse response
            delay: desired equalizer delay (optional)
        Returns: MSE: Mean Squared Error for the designed equalizer
        """
        L = len(h)
        H = self.convMatrix(h,self.N) #(L+N-1)xN matrix - see Chapter 1
        # compute optimum delay based on MSE
        Hp = np.linalg.pinv(H) #Moore-Penrose Pseudo inverse
        #get index of maximum value using argmax, @ for matrix multiply
        opt_delay = np.argmax(np.diag(H @ Hp))
        self.opt_delay = opt_delay #optimized delay

        if delay==None:
            delay=opt_delay
        elif delay >=(L+self.N-1):
            raise ValueError('Given delay is too large delay (should be < L+N-1')

        k0 = delay
        d=np.zeros(self.N+L-1);d[k0]=1 #optimized position of equalizer delay
        self.w=Hp @ d # Least Squares solution, @ for matrix multiply
        MSE=(1-d.T @ H @ Hp @ d) #MSE and err are equivalent,@ for matrix multiply
        return MSE
```

Let us design a zero-forcing equalizer with $N = 14$ taps for the given channel response in Figure 5.5 and equalizer delay k_0 set to 11. The designed equalizer is tested by sending the channel impulse response as the input, $r[k] = h[k]$. The equalizer is expected to compensate for it completely as indicated in Figure 5.6. The overall system response is computed as the convolution of channel impulse response and the response of the zero-forcing equalizer.

```
# Equalizer Design Parameters
N = 14 # Desired number of taps for equalizer filter
delay = 11

# design zero-forcing equalizer for given channel and get tap weights and
# filter the input through the equalizer find equalizer co-effs for given CIR
from DigiCommPy.equalizers import zeroForcing
zf = zeroForcing(N) #initialize ZF equalizer (object) of length N
mse = zf.design(h=h_k,delay=delay) #design equalizer and get Mean Squared Error
w = zf.w # get the tap coeffs of the designed equalizer filter
# mse = zf.design(h=h_k) # Try this delay optimized equalizer

r_k=h_k # Test the equalizer with the sampled channel response as input
d_k=zf.equalize(r_k) # filter input through the eq
h_sys=zf.equalize(h_k) # overall effect of channel and equalizer
print('ZF equalizer design: N={} Delay={} error={}'.format(N,delay,mse))
print('ZF equalizer weights:{}'.format(w))
```

Compute and plot the frequency response of the channel, equalizer and the overall system. The resulting plot, given in Figure 5.6, shows how the zero-forcing equalizer has compensated the channel response by inverting it.

```
#Frequency response of channel,equalizer & overall system
from scipy.signal import freqz
Omega_1, H_F =freqz(h_k) # frequency response of channel
Omega_2, W =freqz(w) # frequency response of equalizer
Omega_3, H_sys =freqz(h_sys) # frequency response of overall system

fig, ax = plt.subplots(nrows=1,ncols = 1)
ax.plot(Omega_1/pi,20*log(abs(H_F)/max(abs(H_F))),'g',label='channel')
ax.plot(Omega_2/pi,20*log(abs(W)/max(abs(W))),'r',label='ZF equalizer')
ax.plot(Omega_3/pi,20*log(abs(H_sys)/max(abs(H_sys))),'k',label='overall system')
ax.legend();ax.set_title('Frequency response');
ax.set_ylabel('Magnitude(dB)');
ax.set_xlabel('Normalized frequency(x $\pi$ rad/sample)');
fig.show()
```

5.5 Zero-forcing equalizer

The response of the filter in time domain is plotted in Figure 5.7. The plot indicates that the equalizer has perfectly compensated the input when $r[k] = h[k]$.

```
#Plot equalizer input and output(time-domain response)
fig, (ax1,ax2) = plt.subplots(nrows=2,ncols = 1)
ax1.stem( np.arange(0,len(r_k)), r_k, use_line_collection=True)
ax1.set_title('Equalizer input');
ax1.set_xlabel('Samples');
ax1.set_ylabel('Amplitude');
ax2.stem( np.arange(0,len(d_k)), d_k, use_line_collection=True);
ax2.set_title('Equalizer output- N=={} Delay={} error={}'.format(N,delay,mse));
ax2.set_xlabel('Samples');
ax2.set_ylabel('Amplitude');
fig.show()
```

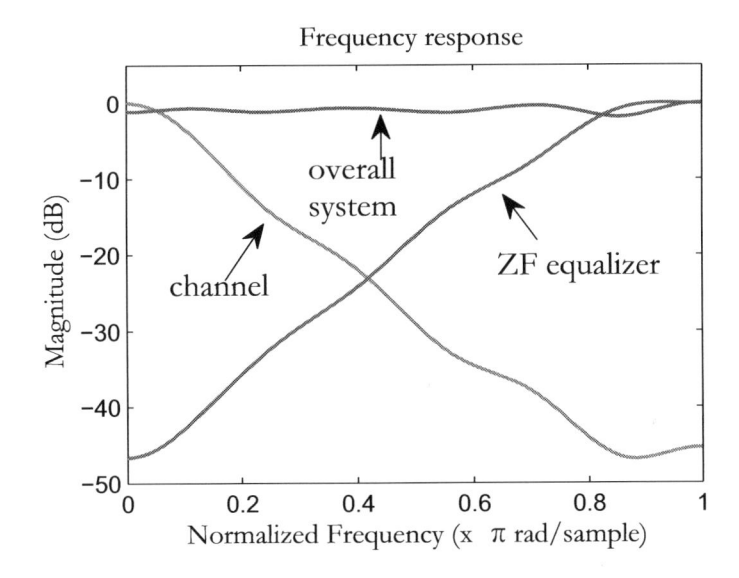

Fig. 5.6: Zero-forcing equalizer compensates for the channel response

The calculated zero-forcing design error depends on the equalizer delay. For an equalizer designed with 14 taps and a delay of 11, the MSE error $=7.04 \times 10^{-4}$ for a noiseless input. On the other hand the error=0.9985 for delay=1. Thus the equalizer delay plays a vital role in the zero-forcing equalizer design. Knowledge of the delay introduced by the equalizer is very essential to determine the position of the first valid sample at the output of the equalizer (see sections 5.7 and 5.8).

The complete simulation code, that includes all the above discussed code snippets, is given next.

Program 69: DigiCommPy\chapter_5\zf_equalizer_test.py: Simulation of zero-forcing equalizer

```
import numpy as np #for numerical computing
import matplotlib.pyplot as plt #for plotting functions
from numpy import pi,log,convolve
nSamp=5 #%Number of samples per symbol determines baud rate Tsym
Fs=100 # Sampling Frequency of the system
```

```python
Ts=1/Fs # Sampling time
Tsym=nSamp*Ts # symbol time period

#Define transfer function of the channel
k=6 # define limits for computing channel response
N0 = 0.001 # Standard deviation of AWGN channel noise
t = np.arange(start=-k*Tsym,stop=k*Tsym,step=Ts) # time base defined till +/-kTsym
h_t = 1/(1+(t/Tsym)**2) # channel model, replace with your own model
h_t = h_t + N0*np.random.randn(len(h_t)) # add Noise to the channel response
h_k = h_t[0::nSamp] # downsampling to represent symbol rate sampler
t_inst=t[0::nSamp] # symbol sampling instants

fig, ax = plt.subplots(nrows=1,ncols = 1)
ax.plot(t,h_t,label='continuous-time model');#response at sampling instants
# channel response at symbol sampling instants
ax.stem(t_inst,h_k,'r',label='discrete-time model',use_line_collection=True)
ax.legend();ax.set_title('Channel impulse response');
ax.set_xlabel('Time (s)');ax.set_ylabel('Amplitude');fig.show()

# Equalizer Design Parameters
N = 14 # Desired number of taps for equalizer filter
delay = 11

# design zero-forcing equalizer for given channel and get tap weights and
# filter the input through the equalizer find equalizer co-effs for given CIR
from DigiCommPy.equalizers import zeroForcing
zf = zeroForcing(N) #initialize ZF equalizer (object) of length N
mse = zf.design(h=h_k,delay=delay) #design equalizer and get Mean Squared Error
w = zf.w # get the tap coeffs of the designed equalizer filter
# mse = zf.design(h=h_k) # Try this delay optimized equalizer

r_k=h_k # Test the equalizer with the sampled channel response as input
d_k=zf.equalize(r_k) # filter input through the eq
h_sys=zf.equalize(h_k) # overall effect of channel and equalizer
print('ZF equalizer design: N={} Delay={} error={}'.format(N,delay,mse))
print('ZF equalizer weights:{}'.format(w))

#Frequency response of channel,equalizer & overall system
from scipy.signal import freqz
Omega_1, H_F  =freqz(h_k) # frequency response of channel
Omega_2, W =freqz(w) # frequency response of equalizer
Omega_3, H_sys =freqz(h_sys) # frequency response of overall system

fig, ax = plt.subplots(nrows=1,ncols = 1)
ax.plot(Omega_1/pi,20*log(abs(H_F)/max(abs(H_F))),'g',label='channel')
ax.plot(Omega_2/pi,20*log(abs(W)/max(abs(W))),'r',label='ZF equalizer')
ax.plot(Omega_3/pi,20*log(abs(H_sys)/max(abs(H_sys))),'k',label='overall system')
ax.legend();ax.set_title('Frequency response');
ax.set_ylabel('Magnitude(dB)');
ax.set_xlabel('Normalized frequency(x $\pi$ rad/sample)');
```

5.5 Zero-forcing equalizer

```
fig.show()

#Plot equalizer input and output(time-domain response)
fig, (ax1,ax2) = plt.subplots(nrows=2,ncols = 1)
ax1.stem( np.arange(0,len(r_k)), r_k, use_line_collection=True)
ax1.set_title('Equalizer input');
ax1.set_xlabel('Samples');
ax1.set_ylabel('Amplitude');
ax2.stem( np.arange(0,len(d_k)), d_k, use_line_collection=True);
ax2.set_title('Equalizer output- N=={} Delay={} error={}'.format(N,delay,mse));
ax2.set_xlabel('Samples');
ax2.set_ylabel('Amplitude');
fig.show()
```

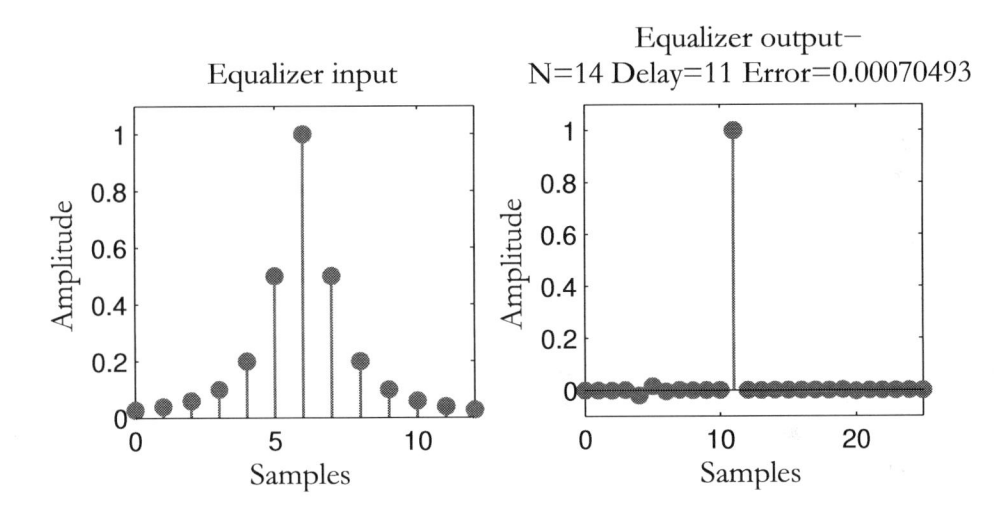

Fig. 5.7: Input samples and the output samples from designed zero-forcing equalizer

5.5.4 Drawbacks of zero-forcing equalizer

Since the response of a zero-forcing equalizer is the inverse of the channel response (refer Figure 5.6), the zero-forcing equalizer applies large gain at those frequencies where the channel has severe attenuation or spectral nulls. As a result, the large gain also amplifies the additive noise at those frequencies and hence the zero-forcing equalization suffers from *noise-enhancement*. Another issue is that the additive white noise (which can be added as part of channel model) can become colored and thereby complicating the optimal detection. The simulated results in Figure 5.8, (obtained by increasing standard deviation of additive noise), show the influence of additive channel noise on the equalizer output. These unwanted effects can be eliminated by applying *minimum mean squared-error* criterion.

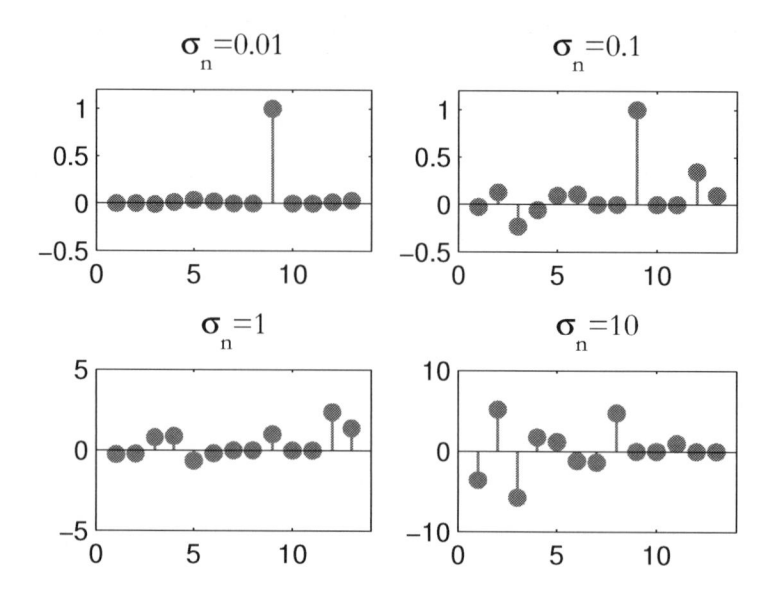

Fig. 5.8: Effect of additive noise on zero-forcing equalizer output

5.6 Minimum mean square error (MMSE) equalizer

The zero-forcing equalizer suffers from the ill-effects of noise enhancement. An improved performance over zero-forcing equalizer can be achieved by considering the noise term in the design of the equalizer. An MMSE filter design achieves this by accounting for a trade-off between noise enhancement and interference suppression.

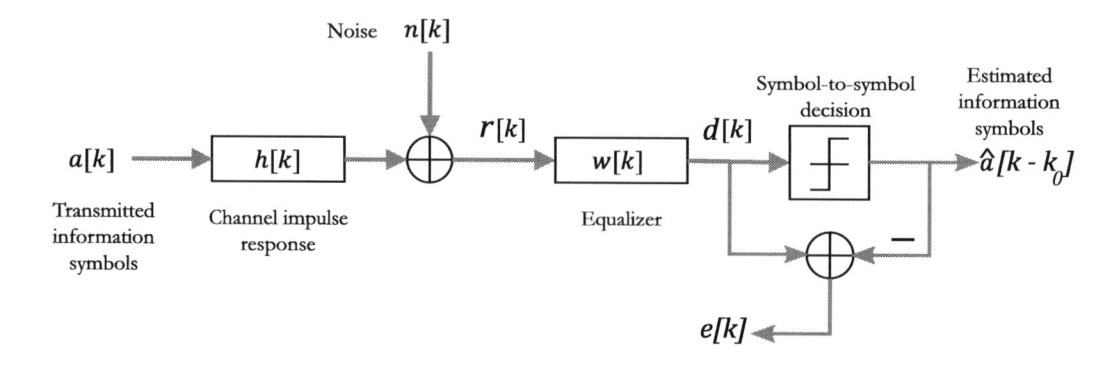

Fig. 5.9: Filter design model for MMSE symbol-spaced linear equalizer

Minimization of error variance and bias is the goal of the MMSE filter design problem, in other words, it tries to minimize the mean square error. In this design problem, we wish to design an FIR MMSE equalizer of length N having the following filter transfer function.

$$W(z) = \sum_{k=0}^{N-1} w[k] z^{-k} \tag{5.26}$$

5.6 Minimum mean square error (MMSE) equalizer

With reference to the generic discrete-time channel model given in Figure 5.9, an MMSE equalizer tries to minimize the variance and bias, from the following error signal

$$e[k] = d[k] - \hat{a}[k - k_0] \tag{5.27}$$

To account for the delay due to the channel and equalizer, the discrete-time model (Figure 5.9) includes the delay k_0 at the decision device output. We note that the error signal depends on the estimated information symbols $\hat{a}[k - k_0]$. Since there is a possibility for erroneous decisions at the output of decision device, it is difficult to account for those errors. Hence, for filter optimization, in training phase, we assume $\hat{a}[k - k_0] = a[k - k_0]$.

$$e[k] = d[k] - a[k - k_0] \tag{5.28}$$

The error signal can be rewritten to include the effect of an N length FIR equalizer on the received symbols

$$e[k] = d[k] - a[k - k_0] = \sum_{i=0}^{N-1} w[i] r[k - i] - a[k - k_0] \tag{5.29}$$

Given the received sequence or noisy measurement $\mathbf{r}$, the goal is to design an equalizer filter $\mathbf{w}$ that would recover the information symbols $\mathbf{a}$. This is a classic *Weiner filter* problem. The Weiner filter design approach requires that we find the filter coefficients $\mathbf{w}$ that minimizes the mean square error ξ. This criterion can be concisely stated as

$$\mathbf{w} = arg\min (\xi) = arg\min E\left\{|e[k]|^2\right\} \tag{5.30}$$

where $E[\cdot]$ is the expectation operator.

For complex signal case, the error signal given in equation 5.29 can be rewritten as

$$\begin{aligned}
e[k] = d[k] - a[k - k_d] &= \sum_{i=0}^{N-1} w^*[i] r[k - i] - a[k - k_0] \\
&= \begin{bmatrix} w_0^* & w_1^* & \cdots & w_{N-1}^* \end{bmatrix} \begin{bmatrix} r_k \\ r_{k-1} \\ \vdots \\ r_{k-(N-1)} \end{bmatrix} - a[k - k_0] \\
&= \mathbf{w}^H \mathbf{r} - a[k - k_0]
\end{aligned} \tag{5.31}$$

where, $\mathbf{w} = [w_0, w_1, \cdots, w_{N-1}]^H$ are the complex filter coefficients with superscript H denoting Hermitian transpose and $\mathbf{r} = [r_k, r_{k-1}, \cdots, r_{k-N-1}]^T$ denotes the complex received sequence.

Using equation 5.31, the *mean square error (MSE)* can be derived as

$$\begin{aligned}
\xi = E\left\{|e[k]|^2\right\} &= E\left\{e[k] e^*[k]\right\} \\
&= E\left\{\left(\mathbf{w}^H \mathbf{r} - a[k - k_0]\right)\left(\mathbf{w}^H \mathbf{r} - a[k - k_0]\right)^H\right\} \\
&= \mathbf{w}^H \underbrace{E\left\{\mathbf{r}\mathbf{r}^H\right\}}_{\mathbf{R}_{rr}} \mathbf{w} - \mathbf{w}^H \underbrace{E\left\{\mathbf{r} a^*[k - k_0]\right\}}_{\mathbf{R}_{ra}} - E\left\{a[k - k_0]\mathbf{r}^H\right\}\mathbf{w} + E\left\{|a[k - k_0]|^2\right\} \\
&= \mathbf{w}^H \mathbf{R}_{rr} \mathbf{w} - \mathbf{w}^H \mathbf{R}_{ra} - \mathbf{R}_{ra}^H \mathbf{w} + E\left\{|a[k - k_0]|^2\right\}
\end{aligned} \tag{5.32}$$

with *autocorrelation matrix* $\mathbf{R}_{rr} = E\left\{\mathbf{r} \cdot \mathbf{r}^H\right\}$ and *crosscorrelation vector* $\mathbf{R}_{ra} = E\left\{\mathbf{r} a^*[k - k_0]\right\}$.

To find the complex tap weights **w** that provide the minimum MSE, the gradient of ξ is set to zero. The gradient of mean square error is given by

$$\frac{\partial}{\partial \mathbf{w}^*}(\xi) = \frac{\partial}{\partial \mathbf{w}^*}\left(\mathbf{w}^H \mathbf{R}_{rr} \mathbf{w}\right) - \frac{\partial}{\partial \mathbf{w}^*}\left(\mathbf{w}^H \mathbf{R}_{ra}\right) \tag{5.33}$$

$$- \frac{\partial}{\partial \mathbf{w}^*}\left(\mathbf{R}_{ra}^H \mathbf{w}\right) + \frac{\partial}{\partial \mathbf{w}^*}\left(E\left\{|a[k-k_0]|^2\right\}\right) \tag{5.34}$$

$$= \mathbf{R}_{rr}\mathbf{w} - \mathbf{R}_{ra} - 0 + 0 \tag{5.35}$$

Equating the gradient of MSE to zero, the Weiner solution to find the complex filter coefficients of the MMSE filter is given by

$$\mathbf{w} = \mathbf{R}_{rr}^{-1}\mathbf{R}_{ra} \tag{5.36}$$

This equation is called as the *Weiner-Hopf equation* with the autocorrelation matrix $\mathbf{R}_{rr}$ and the cross correlation vector $\mathbf{R}_{ra}$ given by

$$\mathbf{R}_{rr} = E\left\{\mathbf{r}\mathbf{r}^H\right\} = \begin{bmatrix} E\{r_k r_k^*\} & E\{r_k r_{k-1}^*\} & \cdots & E\{r_k r_{k-(N-1)}^*\} \\ E\{r_{k-1} r_k^*\} & E\{r_{k-1} r_{k-1}^*\} & \cdots & E\{r_{k-1} r_{k-(N-1)}^*\} \\ \vdots & \vdots & \ddots & \vdots \\ E\{r_{k-(N-1)} r_k^*\} & E\{r_{k-(N-1)} r_{k-1}^*\} & \cdots & E\{r_{k-(N-1)} r_{k-(N-1)}^*\} \end{bmatrix} \tag{5.37}$$

$$\mathbf{R}_{ra} = E\left\{\mathbf{r}a^*[k-k_0]\right\} = \left[E\{r_k a_{k-k_0}^*\} \; E\{r_{k-1} a_{k-k_0}^*\} \; \cdots \; E\{r_{k-(N-1)} a_{k-k_0}^*\}\right]^T \tag{5.38}$$

5.6.1 Alternate solution

An alternate form of equation to solve for MMSE equalizer taps [6] [7], that allow for end-to-end system delay is given by

$$\mathbf{w} = \left(\mathbf{H}^H \mathbf{H} + \frac{\sigma_n^2}{\sigma_a^2}\mathbf{I}\right)^{-1} \mathbf{H}^H \delta_{k_0} \tag{5.39}$$

Here, **H** is the channel matrix as defined in equation 5.14, $\mathbf{H}^H$ its *Hermitian transpose*, **I** is an identity matrix, the ratio $\frac{\sigma_n^2}{\sigma_a^2}$ is the inverse of signal-to-noise ratio (SNR) and the column vector $\delta_{k_0} = [\cdots, 0, 1, 0, \cdots]^T$ with a 1 at k_0th position. This equation is very similar to the equation 5.17 of the zero-forcing equalizer, except for the signal-to-noise ratio term. Therefore, for a noiseless channel, the solution is identical for the MMSE equalizers and the zero-forcing equalizers. This equation allows for end-to-end system delay. Without the equalizer delay (k_0), it is difficult to compensate for the effective channel delay using a causal FIR equalizer. The solution for the MMSE equalizer depends on the availability and accuracy of the channel matrix **H**, the knowledge of the variances σ_a^2 and σ_n^2 (which is often the hardest challenge), and the invertibility of the matrix $\mathbf{HH}^H + \frac{\sigma_n^2}{\sigma_a^2}\mathbf{I}$. In this section, it is assumed that the receiver possesses the knowledge of the channel matrix **H** and the variances - σ_a^2, σ_n^2, and our goal is to design a delay-optimized MMSE equalizer.

The minimum MSE is given by [6] [7]

$$\xi_{min} = \sigma_a^2 \left(1 - \delta_{k_0}^T \mathbf{H}\left[\mathbf{H}^H \mathbf{H} + \frac{\sigma_n^2}{\sigma_a^2}\mathbf{I}\right]^{-1} \mathbf{H}^H \delta_{k_0}\right) \tag{5.40}$$

5.6 Minimum mean square error (MMSE) equalizer

The optimal-delay that minimizes the MSE is simply the index of the maximum diagonal element of the matrix $\mathbf{H} \left[\mathbf{H}^H \mathbf{H} + \frac{\sigma_n^2}{\sigma_a^2} \mathbf{I} \right]^{-1} \mathbf{H}^H$.

$$optimum\ delay = k_{opt} = \underset{index}{\arg\max} \left[diag \left(\mathbf{H} \left[\mathbf{H}^H \mathbf{H} + \frac{\sigma_n^2}{\sigma_a^2} \mathbf{I} \right]^{-1} \mathbf{H}^H \right) \right] \quad (5.41)$$

5.6.2 Design and simulation of MMSE equalizer

A Python function for the design of delay-optimized MMSE equalizer is given next. For implementation, the MMSE equalizer is implemented as a class *derived* from the Equalizer class that was defined in section 5.4. The design method implements equations 5.39, 5.40 and 5.41.

Program 70: DigiCommPy\equalizers.py: Defining the MMSE equalizer derived class

```python
class MMSEEQ(Equalizer): #Class MMSE Equalizer
    def design(self,h,snr,delay=None): #override method in Equalizer abstract class
        """
        Design a MMSE equalizer for given channel impulse response (CIR) and
        signal to noise ratio (SNR). If the tap delay is not given, a delay
        optimized equalizer is designed
        Parameters:
            h : channel impulse response
            snr: input signal to noise ratio in dB scale
            delay: desired equalizer delay (optional)
        Returns: MSE: Mean Squared Error for the designed equalizer
        """
        L = len(h)
        H=self.convMatrix(h,self.N) #(L+N-1)xN matrix - see Chapter 1
        gamma = 10**(-snr/10) # inverse of SNR
        # compute optimum delay
        opt_delay = np.argmax(np.diag(H @ np.linalg.inv(H.T @ H+gamma *
            np.eye(self.N))@ H.T)) # @ for matrix multiply
        self.opt_delay = opt_delay #optimized delay

        if delay==None:
            delay=opt_delay
        if delay >=(L+self.N-1):
            raise ValueError('Given delay is too large delay (should be < L+N-1')

        k0 = delay
        d=np.zeros(self.N+L-1)
        d[k0]=1 # optimized position of equalizer delay
        # Least Squares solution, @ for matrix multiply
        self.w=np.linalg.inv(H.T @ H+ gamma * np.eye(self.N))@ H.T @ d
        # assume var(a)=1, @ for matrix multiply
```

```
        MSE=(1-d.T @ H @ np.linalg.inv(H.T @ H+gamma * np.eye(self.N)) @ H.T @ d)
        return MSE
```

The simulation methodology to test the zero-forcing equalizer was discussed in the previous section. The same methodology is applied here to test MMSE equalizer. Here, the same channel model with ISI length $L = 13$ is used with slightly increased channel noise ($\sigma_n^2 = 0.1$). The channel response at symbol sampling instants are shown in Figure 5.10.

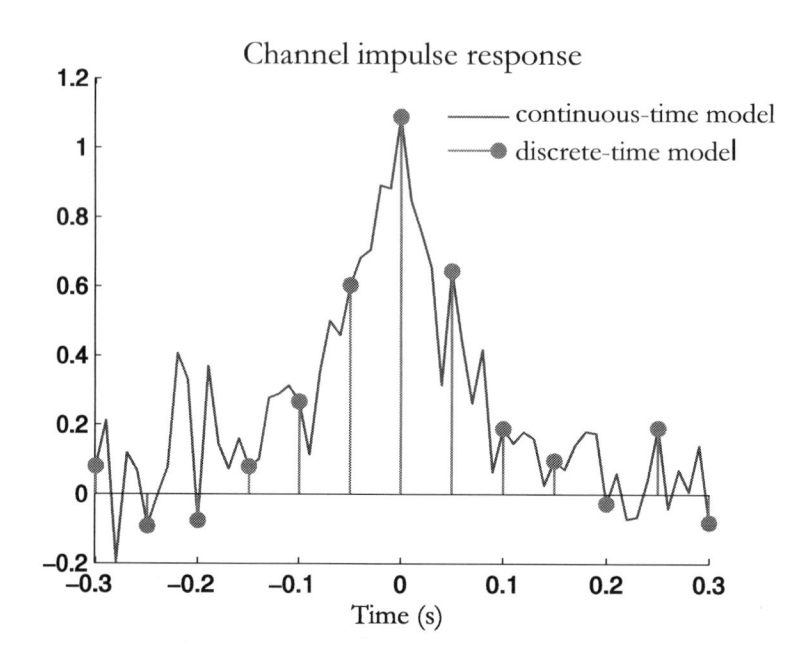

Fig. 5.10: A noisy channel with CIR of length $N = 14$

Next, a delay-optimized MMSE equalizer of length $N = 14$ is designed and used for compensating the distortion introduced by the channel. Figure 5.11 shows the MMSE equalizer in action that beautifully compensates the noisy distorted data. The complete simulation code is given next.

Program 71: DigiCommPy\chapter_5\mmse_equalizer_test.py: Simulation of MMSE equalizer

```python
import numpy as np #for numerical computing
import matplotlib.pyplot as plt #for plotting functions
from numpy import pi,log,log10,convolve
nSamp=5 #%Number of samples per symbol determines baud rate Tsym
Fs=100 # Sampling Frequency of the system
Ts=1/Fs # Sampling time
Tsym=nSamp*Ts # symbol time period

#Define transfer function of the channel
k=6 # define limits for computing channel response
N0 = 0.1 # Standard deviation of AWGN channel noise
t=np.arange(start=-k*Tsym,stop=k*Tsym,step=Ts)#time base defined till +/-kTsym
h_t = 1/(1+(t/Tsym)**2) # channel model, replace with your own model
h_t = h_t + N0*np.random.randn(len(h_t)) # add Noise to the channel response
```

```python
h_k = h_t[0::nSamp] # downsampling to represent symbol rate sampler
t_inst=t[0::nSamp] # symbol sampling instants

fig, ax = plt.subplots(nrows=1,ncols = 1)
ax.plot(t,h_t,label='continuous-time model');#response at all sampling instants
# channel response at symbol sampling instants
ax.stem(t_inst,h_k,'r',label='discrete-time model',use_line_collection=True)
ax.legend();ax.set_title('Channel impulse response');
ax.set_xlabel('Time (s)');ax.set_ylabel('Amplitude');fig.show()

# Equalizer Design Parameters
N = 14 # Desired number of taps for equalizer filter

#design DELAY OPTIMIZED MMSE eq. for given channel, get tap weights and filter
#the input through the equalizer
from DigiCommPy.equalizers import MMSEEQ
noiseVariance = N0**2 # noise variance
snr = 10*log10(1/N0) # convert to SNR (assume var(signal) = 1)
mmse_eq = MMSEEQ(N) #initialize MMSE equalizer (object) of length N
mse = mmse_eq.design(h=h_k,snr=snr)#design equalizer and get Mean Squared Error
w = mmse_eq.w # get the tap coeffs of the designed equalizer filter
opt_delay = mmse_eq.opt_delay

r_k=h_k # Test the equalizer with the sampled channel response as input
d_k=mmse_eq.equalize(r_k) # filter input through the eq
h_sys=mmse_eq.equalize(h_k) # overall effect of channel and equalizer

print('MMSE equalizer design: N={} Delay={} error={}'.format(N,opt_delay,mse))
print('MMSE equalizer weights:{}'.format(w))

#Plot equalizer input and output(time-domain response)
fig, (ax1,ax2) = plt.subplots(nrows=2,ncols = 1)
ax1.stem( np.arange(0,len(r_k)), r_k, use_line_collection=True)
ax1.set_title('Equalizer input');
ax1.set_xlabel('Samples');ax1.set_ylabel('Amplitude');

ax2.stem( np.arange(0,len(d_k)), d_k, use_line_collection=True);
ax2.set_title('Equalizer output- N=={} Delay={} error={}'.format(N,opt_delay,mse));
ax2.set_xlabel('Samples');ax2.set_ylabel('Amplitude');fig.show()
```

5.7 Equalizer delay optimization

For *preset* equalizers, the tap length N and the decision delay k_0 are fixed. The chosen tap length and equalizer delay, significantly affect the performance of the entire communication link. This gives rise to the need for optimizing the tap length and the decision delay. In this section, only the optimization of equalizer delay for both zero-forcing equalizer and MMSE equalizers are simulated. Readers may refer [8] for more details on this topic.

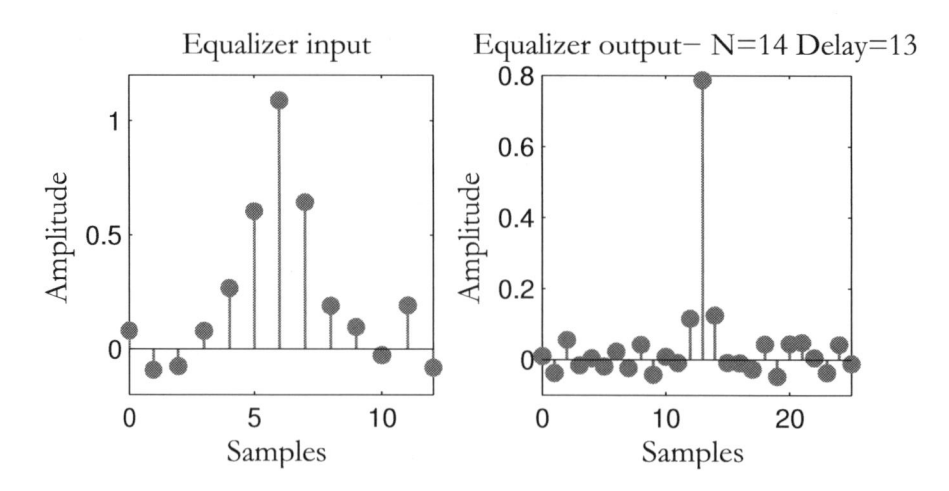

Fig. 5.11: MMSE equalizer of length $L = 14$ and optimized delay $n_0 = 13$

Equalizer delay is very crucial to the performance of the communication link. The equalizer delay determines the value of the symbol that is being detected at the current instant. Therefore, it directly affects the detector block that comes after the equalizer.

The derived classes written in Python (described in the previous sections), for designing zero-forcing equalizer (*zeroForcing*) and the MMSE equalizer (*MMSEQ*), already include the code that calculates the optimum decision delay. The optimization algorithm is based on equations 5.20 and 5.41.

The following code snippet demonstrates the need for delay optimization of an MMSE equalizer for each chosen equalizer tap lengths $N = [5, 10, 15, 20, 25, 30]$. For this simulation, the channel impulse response is chosen as $h[k] = [-0.1, -0.3, 0.4, 1, 0.4, 0.3, -0.1]$. The equalizer delay is swept for each case of tap length and the mean squared error is plotted for visual comparison (Figure 5.12). For a given tap length N, the equalizer with optimum delay is the one that gives the minimum MSE. The same methodology can be applied to test the delay-optimization for the zero-forcing equalizer.

Program 72: DigiCommPy\chapter_5\mmse_eq_delay_opti.py: Delay optimization of MMSE eq.

```python
import numpy as np
import matplotlib.pyplot as plt #for plotting functions

h=np.array([-0.1, -0.3, 0.4, 1, 0.4, 0.3, -0.1]) # test channel
SNR=10 # Signal-to-noise ratio at the equalizer input in dB
Ns= np.arange(start=5, stop=35, step=5) # sweep number of equalizer taps from 5 to 30
maxDelay=Ns[-1]+len(h)-2 #max delay cannot exceed this value
optimalDelay=np.zeros(len(Ns));

from DigiCommPy.equalizers import MMSEEQ

fig, ax = plt.subplots(nrows=1,ncols = 1)
for i,N in enumerate(Ns): #sweep number of equalizer taps
    maxDelay = N+len(h)-2
    mse=np.zeros(maxDelay)
    for j,delay in enumerate(range(0,maxDelay)): # sweep delays
        # compute MSE and optimal delay for each combination
```

5.8 BPSK modulation with zero-forcing and MMSE equalizers

```
            mmse_eq = MMSEEQ(N) #initialize MMSE equalizer (object) of length N
            mse[j]=mmse_eq.design(h,SNR,delay)
            optimalDelay[i] = mmse_eq.opt_delay
        #plot mse in log scale
        ax.plot(np.arange(0,maxDelay),np.log10(mse),label='N='+str(N))
    ax.set_title('MSE Vs eq. delay for given channel and equalizer lengths')
    ax.set_xlabel('Equalizer delay');ax.set_ylabel('$log_{10}$[mse]');
    ax.legend();fig.show()
    #display optimal delays for each selected filter length N. this will correspond
    #with the bottom of the buckets displayed in the plot
    print('Optimal Delays for each N value ->{}'.format(optimalDelay))
```

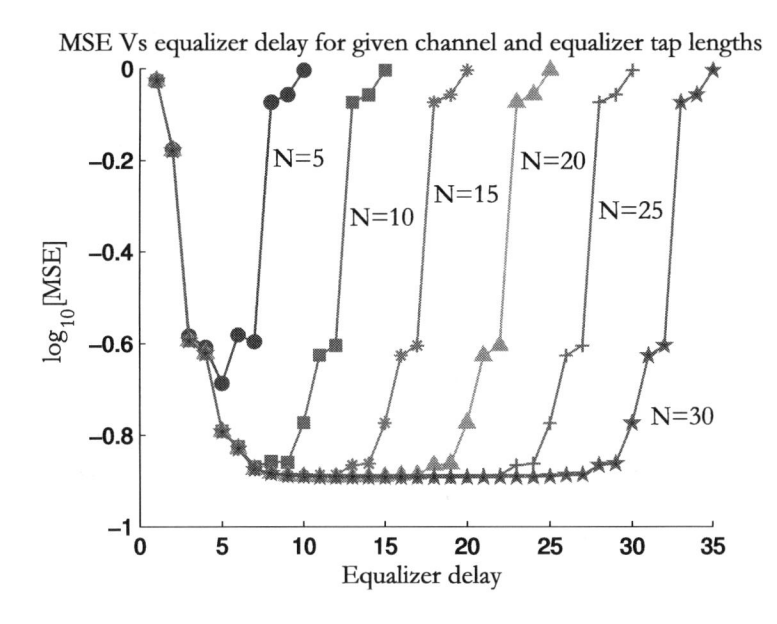

Fig. 5.12: MSE Vs equalizer delay for different values of equalizer tap lengths N

5.8 BPSK modulation with zero-forcing and MMSE equalizers

Having designed and tested the zero-forcing and MMSE FIR linear equalizers, the next obvious task is to test their performance over a communication link. The performance of the linear equalizers completely depend on the characteristics of the channel over which the communication happens. As an example, three distinct channels [4] with the channel characteristics shown in Figure 5.13, are put to test.

In the simulation model, given in Figure 5.14, a string of random data, modulated by BPSK modulation, is individually passed through the discrete channels given in 5.13. In addition to filtering the BPSK modulated data through those channels, the transmitted data is also mixed with additive white Gaussian noise (as dictated by the given E_b/N_0 value).

The received data is then passed through the *delay-optimized* versions of the zero-forcing equalizer and the MMSE equalizer, independently over two different paths. The outputs from the equalizers are passed through

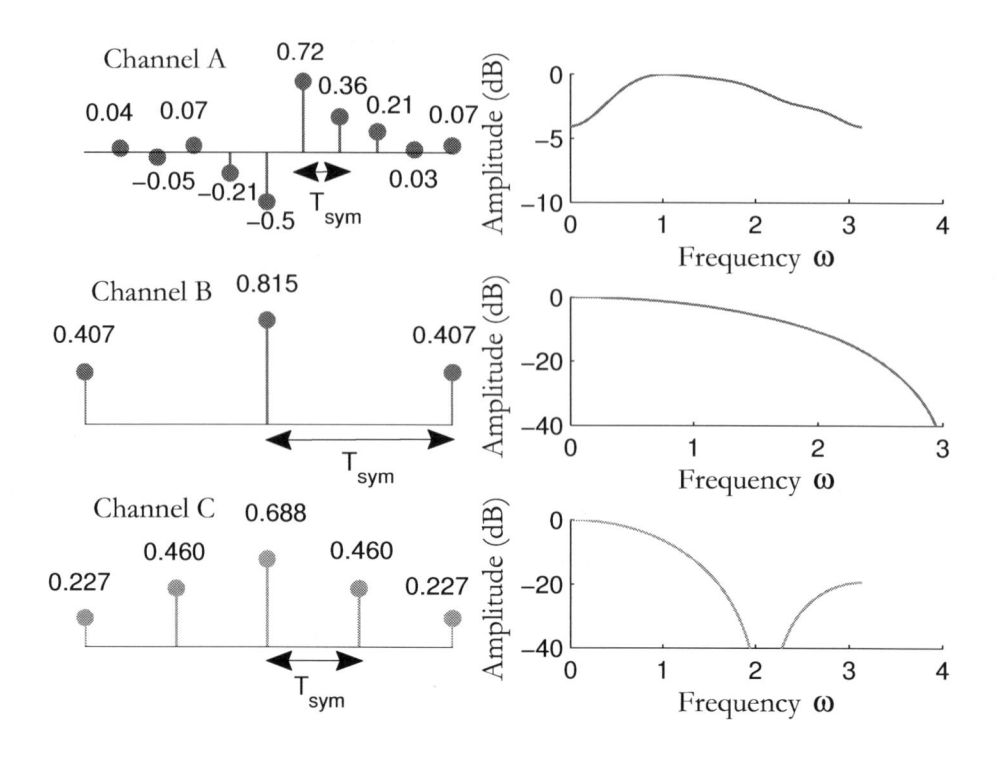

Fig. 5.13: Discrete time characteristics of three channels

the threshold detector that gives an estimate of the transmitted information symbols. The detected symbols are then compared with the original data sequence and the bit-error-rate is calculated. The simulation is performed for different E_b/N_0 values and the resulting performance curves are plotted in Figure 5.15.

The time domain and frequency domain characteristics of the three channels A,B and C are plotted in Figure 5.13. The simulated performance results in Figure 5.15, also include the performance of the communication link over an ISI-free channel (no inter-symbol interference).

The performance of both the equalizers are the best in channel A, which has a typical response of a good telephone channel (channel response is gradual throughout the band with no spectral nulls). The MMSE equalizer performs better than ZF equalizer for channel C, which has the worst spectral characteristic. The next best performance of MMSE equalizer is achieved for channel B. Channel B still has spectral null but it is not as severe as that of channel C. The zero-forcing filters perform the worst over both the channels B and C. This is because the design of zero-forcing filter inherently omits the channel noise in the design equation for the computation of tap weights.

It can be concluded that the linear equalizers yield good performance over well-behaved channels that do not exhibit spectral nulls. On the other hand, they are inadequate for fully compensating inter-symbol interference over the channels that exhibit spectral nulls, which is often the reality. Decision feedback equalizers offer an effective solution for this problem. Non-linear equalizers with reduced complexity are also the focus of researchers to overcome the aforementioned limitations of linear equalizers.

5.8 BPSK modulation with zero-forcing and MMSE equalizers 183

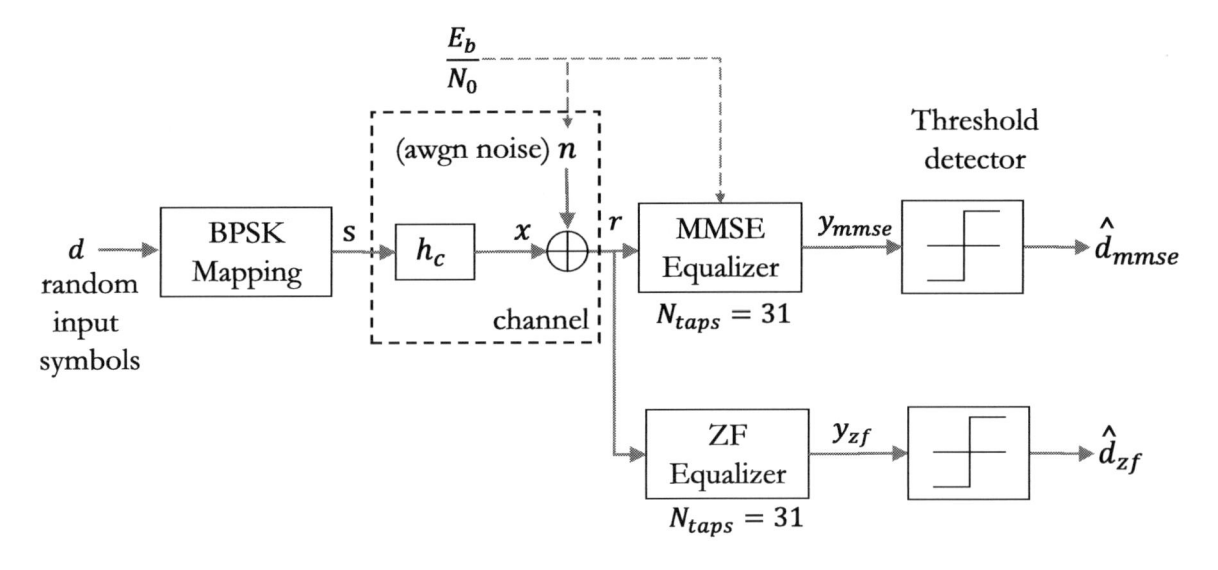

Fig. 5.14: Simulation model for BPSK modulation with zero-forcing and MMSE equalizers

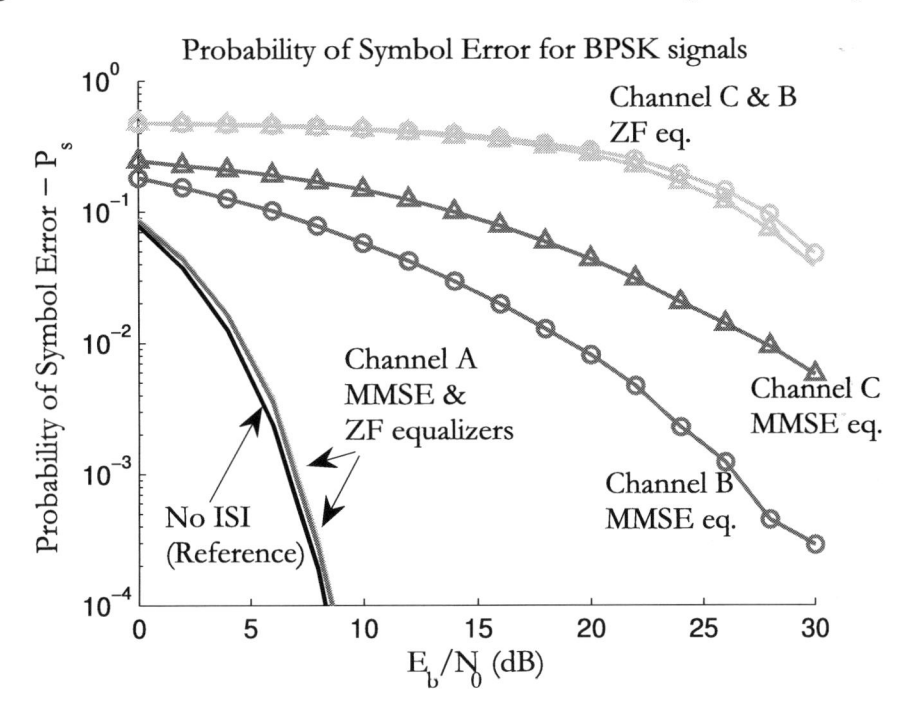

Fig. 5.15: Error rate performance of MMSE and zero-forcing FIR equalizers of length $N = 31$

The complete simulation code is given next. The simulation code reuses the awgn function defined in section 4.1 of chapter 4 and the modem class given in section 3.4.4 of chapter 3.

Program 73: DigiCommPy\chapter_5\isi_equalizers_bpsk.py: Performance of linear equalizers

```python
import matplotlib.pyplot as plt #for plotting functions
from DigiCommPy.modem import PSKModem #import PSKModem
from DigiCommPy.channels import awgn
from DigiCommPy.equalizers import zeroForcing, MMSEEQ #import MMSE equalizer class
from DigiCommPy.errorRates import ser_awgn #for theoretical BERs
from scipy.signal import freqz
#---------Input Fields----------------------
N=10**6 # Number of bits to transmit
EbN0dBs = np.arange(start=0,stop=30,step=2) #  Eb/N0 range in dB for simulation
M=2 # 2-PSK
#h_c=[0.04, -0.05, 0.07, -0.21, -0.5, 0.72, 0.36, 0.21, 0.03, 0.07] # Channel A
#h_c=[0.407, 0.815, 0.407] # uncomment this for Channel B
h_c=[0.227, 0.460, 0.688, 0.460, 0.227] # uncomment this for Channel C
nTaps = 31 # Desired number of taps for equalizer filter
SER_zf = np.zeros(len(EbN0dBs)); SER_mmse = np.zeros(len(EbN0dBs))
#----------------Transmitter--------------------
inputSymbols=np.random.randint(low=0,high=2,size=N) #uniform random symbols 0s & 1s
modem = PSKModem(M)
modulatedSyms = modem.modulate(inputSymbols)
x = np.convolve(modulatedSyms,h_c) # apply channel effect on transmitted symbols

for i,EbN0dB in enumerate(EbN0dBs):
    receivedSyms = awgn(x,EbN0dB) #add awgn noise
    
    # DELAY OPTIMIZED MMSE equalizer
    mmse_eq = MMSEEQ(nTaps) #initialize MMSE equalizer (object) of length nTaps
    mmse_eq.design(h_c,EbN0dB) #Design MMSE equalizer
    optDelay = mmse_eq.opt_delay #get the optimum delay of the equalizer
    #filter received symbols through the designed equalizer
    equalizedSamples = mmse_eq.equalize(receivedSyms)
    y_mmse=equalizedSamples[optDelay:optDelay+N] # samples from optDelay position
    
    # DELAY OPTIMIZED ZF equalizer
    zf_eq = zeroForcing(nTaps) #initialize ZF equalizer (object) of length nTaps
    zf_eq.design(h_c) #Design ZF equalizer
    optDelay = zf_eq.opt_delay #get the optimum delay of the equalizer
    #filter received symbols through the designed equalizer
    equalizedSamples = zf_eq.equalize(receivedSyms)
    y_zf = equalizedSamples[optDelay:optDelay+N] # samples from optDelay position
    
    # Optimum Detection in the receiver - Euclidean distance Method
    estimatedSyms_mmse = modem.demodulate(y_mmse)
    estimatedSyms_zf = modem.demodulate(y_zf)
    # SER when filtered thro MMSE eq.
    SER_mmse[i]=sum((inputSymbols != estimatedSyms_mmse))/N
    # SER when filtered thro ZF eq.
    SER_zf[i]=sum((inputSymbols != estimatedSyms_zf))/N
```

```
SER_theory = ser_awgn(EbN0dBs,'PSK',M=2) #theoretical SER

fig1, ax1 = plt.subplots(nrows=1,ncols = 1)
ax1.semilogy(EbN0dBs,SER_zf,'g',label='ZF Equalizer');
ax1.semilogy(EbN0dBs,SER_mmse,'r',label='MMSE equalizer')
ax1.semilogy(EbN0dBs,SER_theory,'k',label='No interference')
ax1.set_title('Probability of Symbol Error for BPSK signals');
ax1.set_xlabel('$E_b/N_0$(dB)');ax1.set_ylabel('Probability of Symbol Error-$P_s$')
ax1.legend(); ax1.set_ylim(bottom=10**-4, top=1);fig1.show()

# compute and plot channel characteristics
Omega, H_c  = freqz(h_c) #frequency response of the channel
fig2, (ax2,ax3) = plt.subplots(nrows=1,ncols = 2)
ax2.stem(h_c,use_line_collection=True) # time domain
ax3.plot(Omega,20*np.log10(abs(H_c)/max(abs(H_c))));fig2.show()
```

5.9 Adaptive equalizer: Least mean square (LMS) algorithm

To solve for MMSE equalizer coefficients, the Weiner-Hopf solution (equation 5.36) requires the estimation of autocorrelation matrix $\mathbf{R}_{rr}$, correlation matrix $\mathbf{R}_{ra}$ and the matrix inversion $\mathbf{R}_{rr}^{-1}$. This method can be computationally intensive and may render potentially unstable filter.

Alternate solution is to use an iterative approach in which the filter weights $\mathbf{w}$ are updated iteratively in a direction towards the optimum Weiner solution ($\mathbf{R}_{rr}^{-1}\mathbf{R}_{ra}$). This is achieved by iteratively moving the filter weights $\mathbf{w}$ in the direction of the negative gradient

$$\mathbf{w}_{n+1} = \mathbf{w}_n - \mu \nabla \xi[n] \tag{5.42}$$

where, $\mathbf{w}_n$ is the vector of current filter tap weights, $\mathbf{w}_{n+1}$ is the vector of filter tap weights for the update, μ is the step size and $\nabla \xi[n]$ denotes the gradient of the mean square error as derived in equation 5.35.

The *least mean square (LMS) algorithm* is derived from the gradient descent algorithm which is an iterative algorithm to search for the minimum error condition with the cost function equal to the mean square error at output of the equalizer filter. The algorithm is not realizable until we compute $\mathbf{R}_{rr} = E\{\mathbf{rr}^H\}$ and $\mathbf{R}_{ra} = E\{\mathbf{r}a^*[k]\}$, involving an expectation operator. Therefore, for real-time applications, instantaneous squared error is preferred instead of mean squared error. This is achieved by replacing $\mathbf{R}_{rr} = E\{\mathbf{rr}^H\}$ and $\mathbf{R}_{ra} = E\{\mathbf{r}a^*[k]\}$, with instantaneous estimates $\mathbf{R}_{rr} = \mathbf{rr}^H$ and $\mathbf{R}_{ra} = \mathbf{r}a^*[k]$, respectively.

The channel model utilizing LMS algorithm for adapting the FIR linear equalizer is shown in Figure 5.16. The LMS algorithm computes the filter tap updates using the error between the reference sequence and the filter output. The channel model can be operated in two modes: training mode and decision directed mode. In training mode, the transmitted signal $a[k]$ acts as the reference signal for error generation. In decision directed mode (engaged when the actual data transmission happens), the output of the decision device $\hat{a}[k]$ acts as the reference signal for error generation.

With reference to the channel model in Figure 5.16, the recursive equations for computing the error term and the filter coefficient updates of the LMS algorithm in training mode are given by

$$e = a[k] - \mathbf{w}_n^H \mathbf{r} \tag{5.43}$$

$$\mathbf{w}_{n+1} = \mathbf{w}_n + \mu e \mathbf{r} \tag{5.44}$$

where, $\mathbf{w}_n = [w_0, w_1, \cdots, w_{N-1}]^H$ are the current filter coefficients at instant n with superscript H denoting Hermitian transpose and $\mathbf{r} = [r_k, r_{k-1}, \cdots, r_{k-N-1}]^T$ denotes the complex received sequence. For decision directed mode replace $a[k]$ with $\hat{a}[k]$ in the LMS equations.

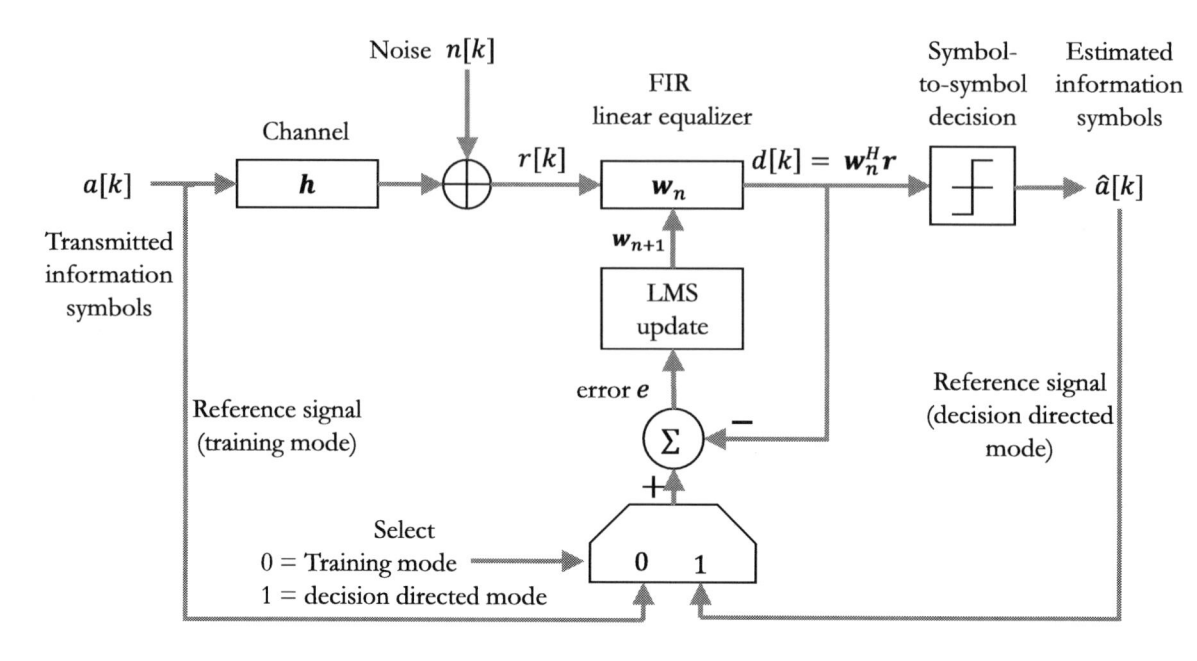

Fig. 5.16: Adaptive filter structure using LMS update

The summary of the LMS algorithm for designing a filter of length N and its Python implementation are given.

Algorithm 1: Least mean square (LMS) algorithm

Input: N : filter length, μ: step size, r : received signal sequence, a: reference sequence
Output: w: final filter coefficients, e: estimation error

1 Initialization: $\mathbf{w}_n = zeros(N,1)$
2 **for** $k = N:length(r)$ **do**
3 $\quad$ Compute $\mathbf{r} = [r_k, r_{k-1}, \cdots, r_{k-N-1}]^T$
4 $\quad$ Compute error term $e = a[k] - \mathbf{w}_n^H \mathbf{r}$
5 $\quad$ Compute filter tap update $\mathbf{w}_{n+1} = \mathbf{w}_n + \mu e \mathbf{r}$
6 **return** w

Program 74: DigiCommPy\equalizers.py: Adaptive equalizer designed using LMS algorithm

```python
class LMSEQ(Equalizer): #Class LMS adaptive equalizer
    def design(self,mu,r,a):
        """
        Design an adaptive FIR filter using LMS update equations (Training Mode)
        Parameters:
            N : desired length of the filter
            mu : step size for the LMS update
            r : received/input sequence
```

```python
        a: reference sequence
    """
    N =self.N
    w = np.zeros(N)
    for k in range(N, len(r)):
        r_vector = r[k:k-N:-1]
        e = a[k] - w @ r_vector.T # @ denotes matrix multiplication
        w = w + mu * e * r_vector # @ denotes matrix multiplication
    self.w = w #set the final filter coefficients
```

To verify the implemented function, a simple test can be conducted. The adaptive filter should be able to identify the response of a short FIR filter whose impulse response is known prior to the test.

Program 75: DigiCommPy\chapter_5\lms_test.py: Verifying the LMS algorithm

```python
import numpy as np
from numpy.random import randn
from numpy import convolve
N = 5 # length of the desired filter
mu=0.1 # step size for LMS algorithm
r=randn(10000) # random input sequence of length 10000
h=randn(N)+1j*randn(N) # random complex system
a=convolve(h,r) # reference signal

from DigiCommPy.equalizers import LMSEQ
lms_eq = LMSEQ(N) #initialize the LMS filter object
lms_eq.design(mu,r,a) # design using input and reference sequences
print('System impulse response (h): {}'.format(h))
print('LMS adapted filter (w): {}'.format(lms_eq.w))
```

A random input signal $r[k]$ and a random complex system impulse response are generated $h[n]$. They are convolved to produce the noiseless reference signal $a[k]$. Pretending that $h[n]$ is unknown, using only the input signal $r[k]$ and the reference signal $a[k]$, the LMS algorithm is invoked to design a FIR filter of length $N = 5$. The resulting optimized filter coefficients $w[n]$ should match the impulse response $h[n]$. Results from a sample run are given next.

```
System impulse response (h):
-0.5726 + 0.0607i 0.0361 - 0.5948i -1.1136 - 0.121i 0.5275 - 0.4212i 1.7004 + 1.8307i
LMS adapted filter (w):
-0.5726 + 0.0607i 0.0361 - 0.5948i -1.1136 - 0.121i 0.5275 - 0.4212i 1.7004 + 1.8307i
```

References

1. J. R. Treichler, I. Fijalkow, and C. R. Johnson, Jr., *Fractionally-spaced equalizers: How long should they really be?*, IEEE Signal Processing Mag., vol. 13, pp. 65–81, May 1996
2. Edward A. Lee, David G. Messerschmitt John R. Barry, *Digital Communications*, 3rd ed.: Springer, 2004
3. R.W. Lucky, *Automatic equalization for digital communication*, Bell System Technical Journal 44, 1965
4. J. G. Proakis, *Digital Communications*, 3rd ed.: McGraw-Hill, 1995.
5. Monson H. Hayes, *Statistical Digital Signal Processing and Modeling*, chapter 4.4.5, Application FIR least squares inverse filter, Wiley, 1 edition, April 11, 1996.

6. C.R Johnson, Hr. et al., *On Fractionally-Spaced Equalizer Design for Digital Microwave Radio Channels*, Signals, Systems and Computers, 1995. 1995 Conference Record of the Twenty-Ninth Asilomar Conference on, vol. 1, pp. 290-294, November 1995.
7. Phil Schniter, *MMSE Equalizer Design*, March 6, 2008, http://www2.ece.ohio-state.edu/~schniter/ee501/handouts/mmse_eq.pdf
8. Yu Gong et al., *Adaptive MMSE Equalizer with Optimum Tap-length and Decision Delay*, Sensor Signal Processing for Defence (SSPD 2010), 29-30 Sept. 2010.

Chapter 6
Receiver Impairments and Compensation

Abstract IQ signal processing is widely used in today's communication receivers. All the IQ processing receiver structures suffer from problems due to amplitude and phase mismatches in their I and Q branches. IQ imbalances are commonplace in analog front-end circuitry and it results in interference of image signal on top of the desired signal. The focus of this chapter is to introduce the reader to some of the basic models of receiver impairments like phase imbalances, gain imbalances and DC offsets. Well known compensation techniques for DC offsets and IQ imbalances are also discussed with code implementation and results.

6.1 Introduction

Direct conversion image rejection receivers [1] for IQ processing - are quite popular in modern RF receiver front-end designs. Direct conversion receivers are preferred over conventional superheterodyne receivers because they do not require separate image filtering [2] [3]. In a direct conversion receiver, the received signal is amplified and filtered at baseband rather than at some intermediate frequency [2]. A direct conversion receiver shown in Figure 6.1 will completely reject the image bands, only if the following two conditions are met : 1) The local oscillator tuned to the desired RF frequency must produce the cosine and sine signals with a phase difference of exactly $90°$, 2) the gain and phase responses of the I and Q branches must match perfectly. In reality, analog components in the RF front-end are not perfect and hence it is impossible to satisfy these requirements. Therefore, complete rejection of the image bands, during RF-IQ conversion, is unavoidable.

The IQ imbalance results from non-ideal RF front-end components. It can also result from power imbalance and non-orthogonality of the I,Q branches due to imperfect local oscillator outputs. In Figure 6.1, the power imbalance on the IQ branches is captured by the gain parameter g on the quadrature branch. The phase mismatch between the local oscillator outputs is captured by the parameter ϕ.

The effect of IQ imbalance is quite disastrous for higher order modulations that form the basis for many modern day communication systems like IEEE 802.11 WLAN, UMTS, LTE, etc. Furthermore, the RF front-end may also introduce DC offsets in the IQ branches, leading to more performance degradation. In order to avoid expensive RF front-end components, signal processing algorithms for compensating the IQ imbalance and DC offsets are a necessity.

In this section, we begin by constructing a model, shown in Figure 6.2 , to represent the effect of following RF receiver impairments

- Phase imbalance and cross-talk on I,Q branches caused by local oscillator phase mismatch - ϕ.
- Gain imbalance on the I,Q branches - g.
- DC offsets in the I and Q branches - dc_i, dc_q.

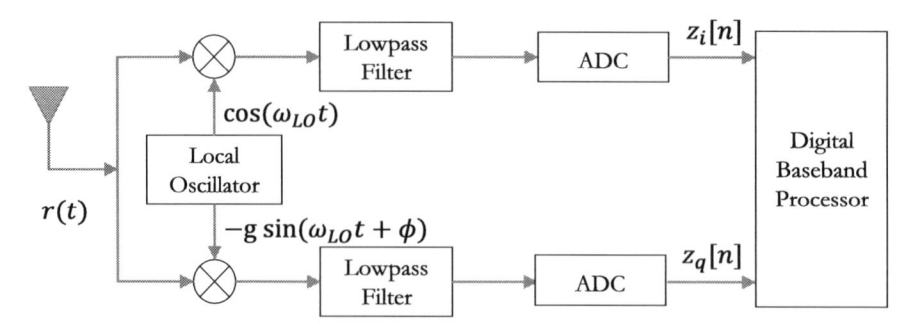

Fig. 6.1: Direct conversion image rejection receiver

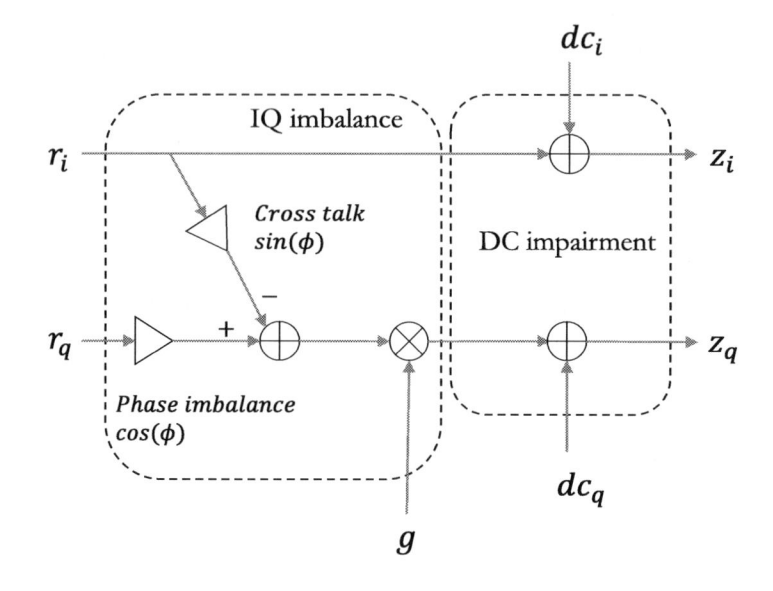

Fig. 6.2: RF receiver impairments model

In this model, the complex signal $r = r_i + jr_q$ denotes the perfect unimpaired signal to the receiver and $z = z_i + jz_q$ represents the signal after the introduction of IQ imbalance and DC offset effects in the receiver front-end.

The following RF receiver impairment model implemented in Python, consists of two parts. First it introduces IQ imbalance in the incoming complex signal and then adds DC offsets to the in-phase and quadrature branches. The details of the individual sub-functions for the IQ imbalance model and DC offset model are described in the subsequent sections. The DC offset impairment model and its compensation is discussed first, followed by the IQ impairment modeling and compensation.

Program 76: DigiCommPy\impairments.py: Class definition for impairments model

```python
import numpy as np
from numpy import sin,cos,pi

class ImpairmentModel():
    # Class: ImpairmentModel
    # Attribute definitions:
```

6.1 Introduction

```
#     self.g : gain mismatch between I,Q branches
#     self.phi : phase mismatch of the local oscillators(in degrees)
#     self.dc_i : DC bias on I branch
#     self.dc_q : DC bias on Q branch
def __init__(self,g=1,phi=0,dc_i=0,dc_q=0): # constructor
    self.g   = g
    self.phi = phi
    self.dc_i = dc_i
    self.dc_q = dc_q

def receiver_impairments(self,r):
    """
    Add receiver impairments to the IQ branches. Introduces DC and IQ
    imbalances between the inphase and quadrature components of the
    complex baseband signal r.
    Parameters:
        r : complex baseband signal sequence to impair
    Returns:
        z : impaired signal sequence
    """
    if isinstance(r,list):
        r = np.array(r) #convert to numpy array if in list format
    k = self.__iq_imbalance(r)
    z = self.__dc_impairment(k)
    return z

def __iq_imbalance(self,r): #private member function
    """
    Add IQ imbalance impairments in a complex baseband. Introduces IQ imbalance
    and phase mismatch between the inphase and quadrature components of the
    complex baseband signal r.
    Parameters:
        r : complex baseband signal sequence to impair
    Returns:
        z : impaired signal sequence
    """
    < See section 6.3 for implementation details >

def __dc_impairment(self,r): #private member function
    """
    Add DC impairments in a complex baseband model.Introduces DC imbalance
    between the inphase and quadrature components of the complex baseband
    signal r.
    Parameters:
        r: complex baseband signal sequence to impair
    Returns:
        z : impaired signal sequence
    """
    < See section 6.2 for implementation details >
```

6.2 DC offsets and compensation

The RF impairment model in the Figure 6.2 contains two types of impairments namely IQ imbalance and DC offset impairment. DC offsets dc_i and dc_q on the I and Q branches are simply modelled as additive factors on the incoming signal.

Program 77: DigiCommPy\impairments.py: Function for adding DC impairments to IQ branches

```
class ImpairmentModel():

    def __init__(self,g=1,phi=0,dc_i=0,dc_q=0): # constructor
        < see section 6.1 for details >

    def receiver_impairments(self,r):
        < see section 6.1 for details >

    def __iq_imbalance(self,r): #private member function
        < see section 6.3 for details >

    def __dc_impairment(self,r): #private member function
        """
        Add DC impairments in a complex baseband model
        Introduces DC imbalance between the inphase and
        quadrature components of the complex baseband signal r.
        Parameters:
            r: complex baseband signal sequence to impair
        Returns:
            z : impaired signal sequence
        """
        return r + (self.dc_i+1j*self.dc_q)
```

Correspondingly, the DC offsets on the branches are simply removed by subtracting the mean of the signal on the I,Q branches from the incoming signal.

Program 78: DigiCommPy\compensation.py: Model for compensating DC offsets in the IQ branches

```
import numpy as np
from numpy import mean,real,imag,sign,abs,sqrt,sum

def dc_compensation(z):
    """
    Function to estimate and remove DC impairments in the IQ branch
    Parameters:
        z: DC impaired signal sequence (numpy format)
    Returns:
        v: DC removed signal sequence
    """
    iDCest=mean(real(z)) # estimated DC on I branch
    qDCest=mean(imag(z)) # estimated DC on I branch
    v=z-(iDCest+1j*qDCest) # remove estimated DCs
    return v
```

6.3 IQ imbalance model

Two IQ imbalance models, namely, *single-branch IQ imbalance model* and *double-branch IQ imbalance model*, are used in practice. These two models differ in the way the IQ mismatch in the inphase and quadrature arms are envisaged. In the single-branch model, the IQ mismatch is modeled as amplitude and phase mismatch in only one of the branches (say Q branch). In contrast, in the double-branch model, the IQ mismatch is modeled as amplitude and phase mismatch in both the I and Q branches. The estimation and compensation algorithms for both these models have to be adjusted accordingly. In this text, only the single-branch IQ imbalance model and its corresponding compensation algorithms are described.

With reference to Figure 6.2, let $r = r_i + jr_q$ denote the perfect unimpaired signal and $z = z_i + jz_q$ represent the signal with IQ imbalance. In this single-branch IQ imbalance model, the gain mismatch is represented by a gain term g in the Q branch only. The difference $1 - g$ is a measure of amplitude deviation of Q branch from the perfectly balanced condition. The presence of phase mismatch ϕ_{rad} in the local oscillator outputs, manifests as cross-talk between the I and Q branches. In essence, together with the gain imbalance g and the phase mismatch ϕ_{rad}, the impaired signals on the I,Q branches are represented as

$$\begin{bmatrix} z_i[k] \\ z_q[k] \end{bmatrix} = \begin{bmatrix} 1 & 0 \\ -g \cdot sin(\phi_{rad}) & g \cdot cos(\phi_{rad}) \end{bmatrix} \begin{bmatrix} r_i[k] \\ r_q[k] \end{bmatrix} \tag{6.1}$$

For the given gain mismatch g and phase mismatch ϕ_{rad} in radians, the following Python function introduces IQ imbalance in a complex baseband signal.

Program 79: DigiCommPy\impairments.py: IQ imbalance model

```python
class ImpairmentModel():

    def __init__(self,g=1,phi=0,dc_i=0,dc_q=0): # constructor
        < see section 6.1 for details >

    def receiver_impairments(self,r):
        < see section 6.1 for details >

    def __iq_imbalance(self,r): #private member function
        """
        Add IQ imbalance impairments in a complex baseband. Introduces IQ imbalance
        and phase mismatch between the inphase and quadrature components of the
        complex baseband signal r.
        Parameters:
            r : complex baseband signal sequence to impair
        Returns:
            z : impaired signal sequence
        """
        Ri=np.real(r); Rq=np.imag(r);
        Zi= Ri # I branch
        Zq= self.g*(-sin(self.phi/180*pi)*Ri \
                + cos(self.phi/180*pi)*Rq) # Q branch crosstalk
        return Zi+1j*Zq

    def __dc_impairment(self,r): #private member function
        < see section 6.2 for details >
```

6.4 IQ imbalance estimation and compensation

Several IQ imbalance compensation schemes are available. This section uses two of them which are based on the works of [4] and [5]. The first algorithm [4] is a blind estimation and compensation algorithm and the one described in [5] is a pilot based estimation and compensation algorithm. Both these techniques are derived for the single-branch IQ imbalance model.

6.4.1 Blind estimation and compensation

The blind technique is a simple, low complexity algorithm, that is solely based on the statistical properties of the incoming complex signal. It does not require additional processing overheads like preamble or training symbols for the IQ imbalance estimation.

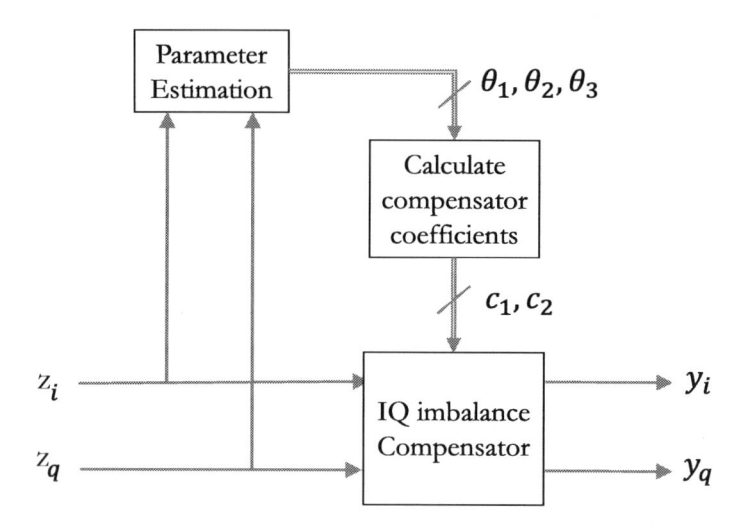

Fig. 6.3: Blind estimation and compensation for IQ imbalance

The blind compensation scheme is shown in Figure 6.3. Let $z = z_i + jz_q$, represent the IQ imbalance-impaired complex baseband signal, that needs to be compensated. First, the complex baseband signal z is used to estimate three imbalance parameters θ_1, θ_2 and θ_3.

$$\theta_1 = -E\left\{sgn(z_i)z_q\right\}$$
$$\theta_2 = E\left\{|z_i|\right\}$$
$$\theta_3 = E\left\{|z_q|\right\} \tag{6.2}$$

where, $sgn(x)$ is the signum function

$$sgn(x) = \begin{cases} -1 & x < 0 \\ 0 & x = 0 \\ 1 & x > 0 \end{cases} \tag{6.3}$$

The compensator co-efficients c_1 and c_2 are then calculated from the estimates of θ_1, θ_2 and θ_3.

6.4 IQ imbalance estimation and compensation

$$c_1 = \frac{\theta_1}{\theta_2} \qquad c_2 = \sqrt{\frac{\theta_3^2 - \theta_1^2}{\theta_2^2}} \qquad (6.4)$$

Finally, the inphase and quadrature components of the compensated signal $y = y_i + jy_q$, are calculated as

$$y_i = z_i \qquad y_q = \frac{c_1 z_i + z_q}{c_2}$$

Program 80: DigiCommPy\compensation.py: Blind IQ imbalance estimation and compensation

```
def blind_iq_compensation(z):
    """
    Function to estimate and compensate IQ impairments for the
    single-branch IQ impairment model
    Parameters:
        z: DC impaired signal sequence (numpy format)
    Returns:
        y: IQ imbalance compensated signal sequence
    """
    I=real(z);Q=imag(z)
    theta1=(-1)*mean(sign(I)*Q)
    theta2=mean(abs(I)); theta3=mean(abs(Q))
    c1=theta1/theta2
    c2=sqrt((theta3**2-theta1**2)/theta2**2)
    return I +1j*(c1*I+Q)/c2
```

6.4.2 Pilot based estimation and compensation

Pilot based estimation algorithms are widely used to estimate various channel properties. The transmitter transmits a pilot sequence, the IQ imbalance of the receiver front end is estimated based on the received sequence. The pilot estimation technique is well suited for wireless applications like WLAN, UMTS and LTE, where a known pilot sequence is transmitted as part of the data communication session.

The low complexity algorithm, proposed in [5], uses a preamble of length $L = 64$ to estimate the gain imbalance (K_{est}) and phase mismatch P_{est}.

$$K_{est} = \sqrt{\frac{\sum_{k=1}^{L} z_q^2[k]}{\sum_{k=1}^{L} z_i^2[k]}} \qquad (6.5)$$

$$P_{est} = \frac{\sum_{k=1}^{L} (z_i[k] \cdot z_q[k])}{\sum_{k=1}^{L} z_i^2[k]} \qquad (6.6)$$

where, the complex signal $z = z_i + jz_q$, represents the impaired version of the long preamble as given in the IEEE 802.11a specification [6].

Program 81: DigiCommPy\compensation.py: IQ imbalance estimation using Pilot transmission

```python
class PilotEstComp():
    # Class: PilotEstComp (Pilot based estimation and compensation)
    # Attribute definitions:
    #     self.impObj: reference to the object implementing the impairment model
    #     self.Kest : estimated gain imbalance
    #     self.Pest : estimated phase mismatch
    #     self.preamble : time domain representation long preamble (IEEE 802.11a)
    def __init__(self,impObj): # constructor
        self.impObj = impObj
        self.Kest = 1
        self.Pest = 0
        # Length 64 - long preamble in IEEE 802.11a (frequency domain representation)
        preamble_freqDom = np.array([0,0,0,0,0,0,1,1,\
                            -1,-1,1,1,-1,1,-1,1,\
                            1,1,1,1,1,-1,-1,1,1,\
                            -1,1,-1,1,1,1,1,0,1,\
                            -1,-1,1,1,-1,1,-1,1,\
                            -1,-1,-1,-1,-1,1,1,\
                            -1,-1,1,-1,1,-1,1,1,\
                            1,1,0,0,0,0,0])
        from scipy.fftpack import ifft
        self.preamble=ifft(preamble_freqDom,n = 64)

    def pilot_est(self):
        """
        IQ imbalance estimation using Pilot transmission
        Computes:
            Kest - estimated gain imbalance
            Pest - estimated phase mismatch
        """
        # send known preamble through the impairments model
        r_preamb = self.impObj.receiver_impairments(self.preamble)

        # remove DC imbalance before IQ imbalance estimation
        z_preamb= r_preamb - (mean(real(r_preamb)) + 1j* mean(imag(r_preamb)))
        # IQ imbalance estimation
        I=real(z_preamb); Q=imag(z_preamb)
        self.Kest = sqrt(sum((Q*Q))/sum(I*I)) # estimated gain imbalance
        self.Pest = sum(I*Q)/sum(I*I) # estimated phase mismatch

    def pilot_iqImb_compensation(self,d):
        < see next program >
```

Let $d = d_i + jd_q$ be the impaired version of complex signal received during the normal data transmission interval. Once the parameters given in equation 6.5 and 6.6 are estimated during the preamble transmission, the IQ compensation during the normal data transmission is as follows

6.5 Visualizing the effect of receiver impairments

$$w_i = d_i$$
$$w_q = \frac{d_q - P_{est} \cdot d_i}{k_{est}\sqrt{1 - P_{est}^2}} \quad (6.7)$$

Program 82: DigiCommPy\compensation.py: Compensation of IQ imbalance during data transmission

```
class PilotEstComp():

    def __init__(self,impObj): # constructor
        < see previous program >

    def pilot_est(self):
        < see previous program >

    def pilot_iqImb_compensation(self,d):
        """
        Function to compensate IQ imbalance during the data transmission
        Parameters:
            d : The impaired received complex signal sequence
        Returns:
            w : IQ imbalance compensated complex signal sequence
        Usage:
            from compensation import PilotEstComp
            pltEstCompObj = PilotEstComp(impObj) #initialize
            pltEstCompObj.pilot_iqImb_compensation(d) #call function
        """
        # estimate the Kest, Pest for the given model using pilot transmission
        self.pilot_est()
        d_dcRemoved = d - (mean(real(d)) + 1j* mean(imag(d)))
        I=real(d_dcRemoved); Q=imag(d_dcRemoved)
        wi= I;
        wq = (Q - self.Pest*I)/sqrt(1-self.Pest**2)/self.Kest
        return wi + 1j*wq
```

6.5 Visualizing the effect of receiver impairments

With the receiver impairments model and compensation techniques in place, let us visualize their effects in a complex plane. For instance, consider a higher order modulation like 64-QAM. The receiver impairments like IQ imbalance and DC offsets are added to the QAM modulated signal. The unimpaired sequence and the sequence affected by receiver impairments are plotted on a complex plane as shown in Figure 6.4. The following code re-uses the QAMModem class that was already defined in section 3.4.3 of chapter 3.

Program 83: DigiCommPy\chapter_6\rf_impairments.py: Visualizing receiver impairments

```
import numpy as np #for numerical computing
from numpy import real,imag
from DigiCommPy.modem import QAMModem #QAM Modem model
from DigiCommPy.impairments import ImpairmentModel #Impairment Model
```

```python
import matplotlib.pyplot as plt #for plotting functions

M=64 # M-QAM modulation order
nSym=1000 # To generate random symbols

# uniform random symbols from 0 to M-1
inputSyms = np.random.randint(low=0, high = M, size=nSym)
modem = QAMModem(M) #initialize the M-QAM modem object
s = modem.modulate(inputSyms) #modulated sequence

impModel_1 =  ImpairmentModel(g=0.8) # gain mismatch only model
impModel_2 =  ImpairmentModel(phi=12) # phase mismatch only model
impModel_3 =  ImpairmentModel(dc_i=0.5,dc_q=0.5) # DC offsets only
impModel_4 =  ImpairmentModel(g=0.8,phi=12,dc_i=0.5,dc_q=0.5) # All impairments

#Add impairments to the input signal sequence using the models
r1 = impModel_1.receiver_impairments(s)
r2 = impModel_2.receiver_impairments(s)
r3 = impModel_3.receiver_impairments(s)
r4 = impModel_4.receiver_impairments(s)

fig, ax = plt.subplots(nrows=2,ncols = 2)

ax[0,0].plot(real(s),imag(s),'b.')
ax[0,0].plot(real(r1),imag(r1),'r.');ax[0,0].set_title('IQ Gain mismatch only')

ax[0,1].plot(real(s),imag(s),'b.')
ax[0,1].plot(real(r3),imag(r3),'r.');ax[0,1].set_title('DC offsets only')

ax[1,0].plot(real(s),imag(s),'b.')
ax[1,0].plot(real(r2),imag(r2),'r.');ax[1,0].set_title('IQ Phase mismatch only')

ax[1,1].plot(real(s),imag(s),'b.')
ax[1,1].plot(real(r4),imag(r4),'r.')
ax[1,1].set_title('IQ impairments & DC offsets');fig.show()
```

6.6 Performance of M-QAM modulation with receiver impairments

In a wireless communication system, that uses OFDM with higher order modulations like M-QAM, the effect of receiver impairments is of great concern. The performance of a higher order M-QAM under different conditions of receiver impairments is simulated here. Figure 6.5 shows the functional blocks used in this performance simulation. A received M-QAM signal is impaired by IQ imbalance (both gain imbalance g and phase imbalance ϕ) and DC offsets (dc_i, dc_q) on the IQ arms. First, the DC offset is estimated and compensated; Next, the IQ compensation is separately performed using blind IQ compensation and pilot based IQ compensation. The symbol error rates are finally computed for following four cases: (a) received signal (with no compensation), (b) with DC compensation alone, (c) with DC compensation and blind IQ compensation and (d) with DC compensation and pilot based IQ compensation.

6.6 Performance of M-QAM modulation with receiver impairments

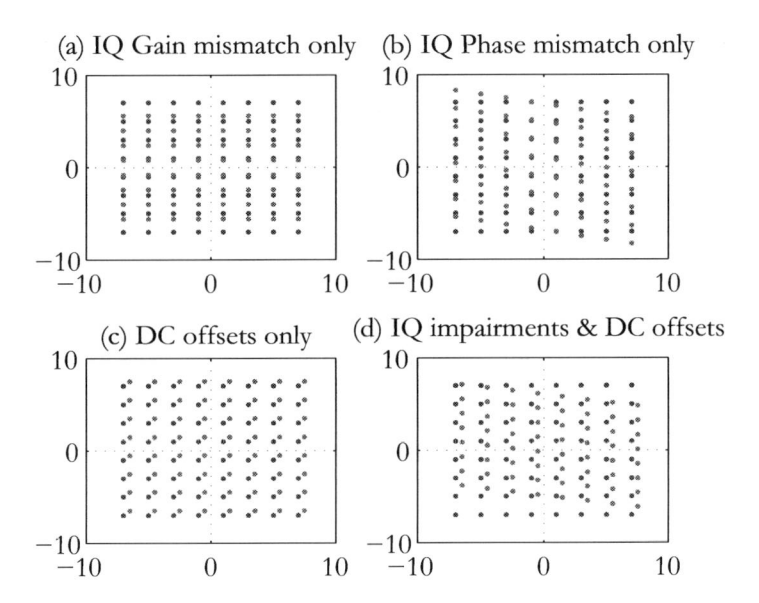

Fig. 6.4: Constellation plots for unimpaired (blue) and impaired (red) versions of 64-QAM modulated symbols: (a) Gain imbalance $g = 0.8$ (b) Phase mismatch $\phi = 12°$ (c) DC offsets $dc_i = 0.5, dc_q = 0.5$ (d) With all impairments $g = 0.8, \phi = 12°, dc_i = 0.5, dc_q = 0.5$

The complete simulation code for this performance simulation is given next. The code re-uses many functions defined in chapter 3 and chapter 4. The function to perform MQAM modulation and the coherent detection technique was already defined in sections 3.4.3 and 3.4.4. The AWGN noise model and the code to compute theoretical symbol error rates are given in sections 4.1.2 and 4.1.3 respectively.

Figure 6.6 illustrates the results obtained for the following values of receiver impairments: $g = 0.9, \phi = 8°, dc_i = 1.9, dc_q = 1.7$. For this condition, both the blind IQ compensation and pilot based IQ compensation techniques are on par. Figure 6.7 illustrates the results obtained for the following values of receiver impairments: $g = 0.9, \phi = 30°, dc_i = 1.9, dc_q = 1.7$, where the phase mismatch is drastically increased compared to the previous case. For this condition, the performance with the pilot based IQ compensation is better than that of the blind compensation technique. Additionally, the constellation plots of the impaired signal and the compensated signal, are also plotted for various combinations of DC offsets, gain imbalances and phase imbalances, which are shown in Figure 6.8.

Program 84: DigiCommPy\chapter_6\mqam_awgn_iq_imb.py: Performance of M-QAM modulation technique with receiver impairments

```python
import numpy as np # for numerical computing
from DigiCommPy.modem import QAMModem #QAM Modem model
from DigiCommPy.channels import awgn
from DigiCommPy.impairments import ImpairmentModel #Impairment Model
from DigiCommPy.compensation import 
    dc_compensation,blind_iq_compensation,PilotEstComp
from DigiCommPy.errorRates import ser_awgn
import matplotlib.pyplot as plt #for plotting functions

# ---------Input Fields-----------------------
nSym=100000 # Number of input symbols
```

```python
EbN0dBs = np.arange(start=-4, stop=24, step=2) # Define EbN0dB range for simulation
M=64 # M-QAM modulation order
g=0.9; phi=8; dc_i=1.9; dc_q=1.7 # receiver impairments
# --------------------------------------------
k=np.log2(M)
EsN0dBs = 10*np.log10(k)+EbN0dBs # EsN0dB calculation

SER_1 = np.zeros(len(EbN0dBs)) # Symbol Error rates (No compensation)
SER_2 = np.zeros(len(EbN0dBs)) # Symbol Error rates (DC compensation only)
SER_3 = np.zeros(len(EbN0dBs)) # Symbol Error rates (DC comp & Blind IQ comp)
SER_4 = np.zeros(len(EbN0dBs)) # Symbol Error rates (DC comp & Pilot IQ comp)

d = np.random.randint(low=0, high = M, size=nSym) # random symbols from 0 to M-1
modem = QAMModem(M) #initialize the M-QAM modem object
modulatedSyms = modem.modulate(d) #modulated sequence

for i,EsN0dB in enumerate(EsN0dBs):
    receivedSyms = awgn(modulatedSyms,EsN0dB) # add awgn nois
    impObj =  ImpairmentModel(g,phi,dc_i,dc_q) # init impairments model
    y1 = impObj.receiver_impairments(receivedSyms) # add impairments
    
    y2 = dc_compensation(y1) # DC compensation
    #Through Blind IQ compensation after DC compensation
    y3 = blind_iq_compensation(y2)
    #Through Pilot estimation and compensation model
    pltEstCompObj = PilotEstComp(impObj) #initialize
    y4 = pltEstCompObj.pilot_iqImb_compensation(y1) #call function
    """
    Enable this section - if you want to plot constellation diagram
    fig1, ax = plt.subplots(nrows = 1,ncols = 1)
    ax.plot(np.real(y1),np.imag(y1),'r.')
    ax.plot(np.real(y4),np.imag(y4),'b*')
    ax.set_title('$E_b/N_0$={} (dB)'.format(EbN0dBs[i]));
    fig1.show();input()
    """
    # -------IQ Detectors--------
    dcap_1 = modem.iqDetector(y1) # No compensation
    dcap_2 = modem.iqDetector(y2) # DC compensation only
    dcap_3 = modem.iqDetector(y3) # DC & blind IQ comp.
    dcap_4 = modem.iqDetector(y4) # DC & pilot IQ comp.
    
    # ------ Symbol Error Rate Computation-------
    SER_1[i]=sum((d!=dcap_1))/nSym; SER_2[i]=sum((d!=dcap_2))/nSym
    SER_3[i]=sum((d!=dcap_3))/nSym; SER_4[i]=sum((d!=dcap_4))/nSym

SER_theory = ser_awgn(EbN0dBs,'QAM',M) #theory SER
fig2, ax = plt.subplots(nrows = 1,ncols = 1)
ax.semilogy(EbN0dBs,SER_1,'*-r',label='No compensation')
ax.semilogy(EbN0dBs,SER_2,'o-b',label='DC comp only')
ax.semilogy(EbN0dBs,SER_3,'x-g',label='Sim- DC and blind iq comp')
```

6.6 Performance of M-QAM modulation with receiver impairments

```
ax.semilogy(EbN0dBs,SER_4,'D-m',label='Sim- DC and pilot iq comp')
#ax.semilogy(EbN0dBs,SER_theory,'k',label='Theoretical')
ax.set_xlabel('$E_b/N_0$ (dB)');ax.set_ylabel('Symbol Error Rate ($P_s$)')
ax.set_title('Probability of Symbol Error 64-QAM signals');
ax.legend();fig2.show()
```

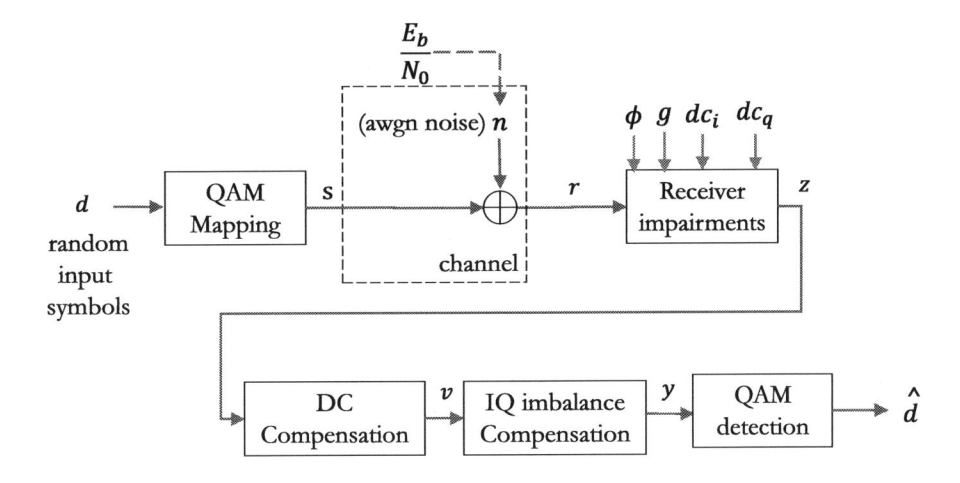

Fig. 6.5: Simulation model for M-QAM system with receiver impairments

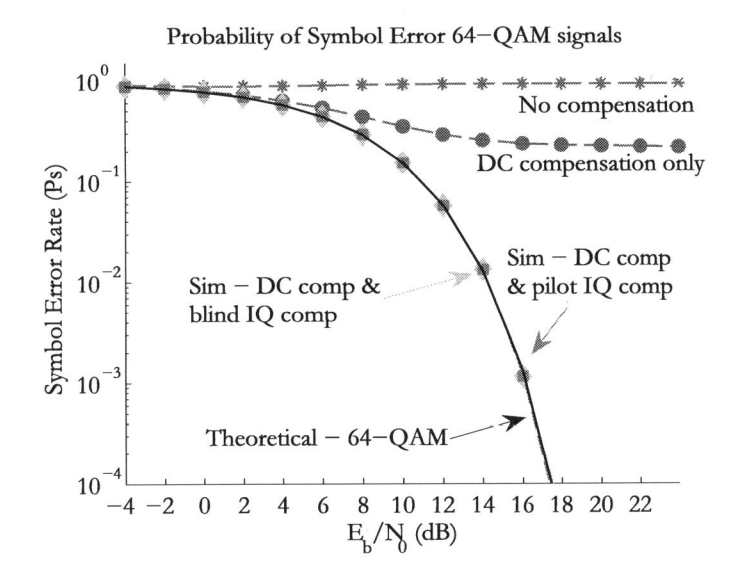

Fig. 6.6: Performance of 64-QAM with receiver impairments: $g = 0.9, \phi = 8°, dc_i = 1.9, dc_q = 1.7$

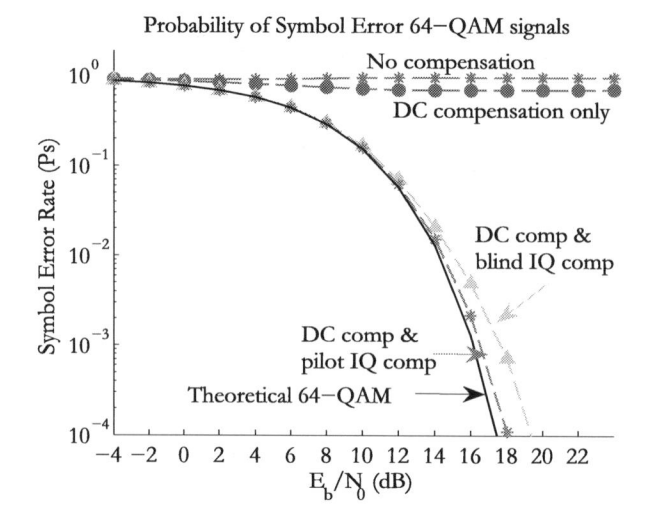

Fig. 6.7: Performance of 64-QAM with receiver impairments: $g = 0.9, \phi = 30°, dc_i = 1.9, dc_q = 1.7$

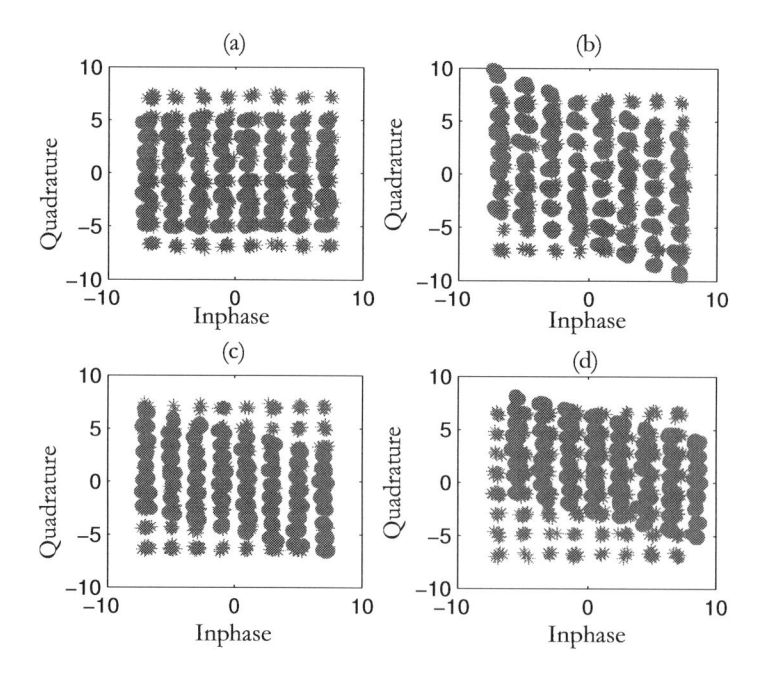

Fig. 6.8: Constellation plots at $E_b/N_0 = 21 dB$ for various combinations of receiver impairments: (a) Gain imbalance only - $g = 0.7$ (b) Phase mismatch only - $\phi = 25°$ (c) Gain imbalance and phase mismatch $g = 0.7, \phi = 25°$ (d) All impairments - $g = 0.7, \phi = 25°, dc_i = 1.5, dc_q = 1.5$

References

1. T.H. Meng W. Namgoong, *Direct-conversion RF receiver design*, IEEE Transaction on Communications, 49(3):518–529, March 2001
2. A. Abidi, *Direct-conversion radio transceivers for digital communications*, IEEE Journal of Solid-State Circuits, 30:1399–1410, December 1995.

References

3. B. Razavi, RF microelectronics, ISBN 978-0137134731, Prentice Hall, 2 edition, October 2011
4. Moseley, Niels A., and Cornelis H. Slump. *A low-complexity feed-forward I/Q imbalance compensation algorithm*, 17th Annual Workshop on Circuits, 23-24 Nov 2006, Veldhoven, The Netherlands. pp. 158-164. Technology Foundation STW. ISBN 978-90-73461-44-4
5. K.H Lin et al., *Implementation of Digital IQ Imbalance Compensation in OFDM WLAN Receivers*, IEEE International Symposium on Circuits and Systems, 2006
6. *Wireless LAN Medium Access Control (MAC) and Physical Layer (PHY) Specifications: High-Speed Physical Layer in the 5 GHz Band*, IEEE Standard 802.11a-1999-Part II, Sep.1999.

Index

adaptive equalizer, 160, 185
additive white Gaussian noise (AWGN), 139
analytic signal, 37
angle of arrival (AoA), 154
angular frequency, 42
autocorrelation matrix, 175

block fading, 148
bounded input bounded output (BIBO), 50

causal filter, 49
channel impulse response (CIR), 147
circular convolution, 33, 35
coherent BFSK, 109, 132
coherent detection, 130
complex baseband equivalent representation, 122
complex baseband models
    channel models
        AWGN channel, 139
        linear time invariant (LTI), 147
        Rayleigh flat fading, 149
        Rician flat fading, 154
    demodulators
        IQ Detector, 130
        M-FSK, 133
    modulators
        MFSK, 132
        MPAM, 125
        MPSK, 126
        MQAM, 127
complex envelope, 122
continuous phase frequency shift keying (CPFSK), 85
continuous phase modulation (CPM), 85
continuous time fourier transform (CTFT), 37
convolution, 30
    circular, 33, 35
    linear, 31
crosscorrelation vector, 175

dc compensation, 192
discrete-time fourier transform (DTFT), 38

energy of a signal, 25
envelope, 42

envelope detector, 133
equalizer, 159
    adaptive, 160
    decision feedback, 160
    fractionally-spaced, 161
    linear, 160
    maximum likelihood sequence estimation (MLSE), 160
    minimum mean square error (MMSE), 174
    preset, 160
    symbol-spaced, 160
Euclidean distance, 166
Euclidean norm, 25, 26
Eucliden norm, 148

fading, 154
Fast Fourier Transform, 12, 35
    FFTshift, 15
    IFFTShift, 17
finite impulse response (FIR), 47, 147, 165
flat fading channel, 146, 148
fractionally-spaced equalizer, 161
frequency selective channel, 146
frequency-non-selective channel, 148

Gibbs phenomenon, 4
group delay, 52

Hermitian transpose, 166, 175, 186
Hilbert transform, 38

infinite impulse response (IIR), 47
instantaneous amplitude, 42
instantaneous phase, 42
IQ detection, 130
IQ imbalance model, 193

least mean square (LMS) algorithm, 185
least squares (LS), 166
line of sight (LOS) path, 154
linear convolution, 31
linear equalizers, 159
linear phase filter, 52
linear time invariant (LTI) channel, 147

maximum likelihood sequence estimation (MLSE), 86
mean square error (MSE), 167, 175
mean square error criterion, 161
minimum mean square error (MMSE) equalizer, 174
minimum shift keying (MSK), 86, 87, 111
modulation index, 87, 109
modulation order, 125
modulation with memory, 85
moment generating function (MGF), 150, 155
Moore-Penrose generalized matrix inverse, 167

non line of sight (NLOS) path, 154
non-causal filter, 49
non-coherent BFSK, 109
non-coherent detection, 130
norm, 29
    2-norm, 25, 29
    L2 norm, 25
    p-norm, 29
Nyquist sampling theorem, 1

passband model, 55
peak distortion criterion, 161
periodogram, 24
phase delay, 52
phase distortion, 52
pilot based estimation, 195
polynomial functions, 30
power of a signal, 26
power spectral density (PSD), 23, 98
preset equalizers, 160
pseudo-inverse matrix, 167

raised cosine pulse, 94
Rayleigh flat-fading, 149
RF impairments model, 189
Rician flat-fading, 154

Rician K factor, 154

signal
    energy signal, 27
    power signal, 27
signum function, 4
symbol-spaced equalizer, 160

tapped delay line (TDL) filter, 147
temporal fine structure (TFS), 43
temporal frequency, 42
test signals, 1
    chirp signal, 7
    gaussian pulse, 6
    rectangular pulse, 5
    sinusoidal, 1
    square wave, 3
Toeplitz matrix, 33, 166

waveform simulation models
    BFSK, 109
        coherent, 111
        non-coherent, 111
    BPSK, 56
    D-BPSK, 62
    DEBPSK, 59
    GMSK, 100
    MSK, 87
    O-QPSK, 73
    pi/4-QPSK, 79
    QPSK, 67
Weiner filter, 175
Weiner-Hopf equation, 176
Wiener-Khinchin theorem, 23

zero-forcing equalizer, 165

Printed in Great Britain
by Amazon